Rebecca Mertens
The Construction of Analogy-Based Research Programs

Science Studies

Rebecca Mertens (PhD), born in 1984, is a postdoctoral researcher in the history and philosophy of science at the University of Bielefeld, Germany. She works on the role of analogies, models and forms of comparison in the history of molecular genetics and is a member of the collaborative research program "Practices of Comparison: Ordering and Changing the World". During her graduate and doctoral studies, she was a visiting scholar at the École Normale Supérieure in Paris and a visiting graduate fellow at the Minnesota Center for Philosophy of Science.

Rebecca Mertens
The Construction of Analogy-Based Research Programs
The Lock-and-Key Analogy in 20th Century Biochemistry

[transcript]

This book has been submitted as a dissertation to the University of Bielefeld.

Bibliographic information published by the Deutsche Nationalbibliothek
The Deutsche Nationalbibliothek lists this publication in the Deutsche Nationalbibliografie; detailed bibliographic data are available in the Internet at http://dnb.d-nb.de

© 2019 transcript Verlag, Bielefeld

Cover layout: Maria Arndt, Bielefeld
Cover illustration: Myoglobin 3D structure, AzaToth (https://de.wikipedia.org/
    wiki/ProteinÞ/media/File:Myoglobin.png)
Printed by Majuskel Medienproduktion GmbH, Wetzlar
Print-ISBN 978-3-8376-4442-5
PDF-ISBN 978-3-8394-4442-9
https://doi.org/10.14361/9783839444429

Table of contents

Preface | 7

1 The lock-and-key analogy and its influence on 20th century biochemistry | 9
1.1 State of the literature | 14
1.2 Theoretical approach: The philosophical analysis of analogies in science | 21
1.3 Methodology | 29
1.4 Outline | 33
1.5 Sources | 36

2 The lock-and-key analogy in Emil Fischer's program on sugar fermentation, 1890-1907 | 39
2.1 Origins of the concept of molecular geometry and Fischer's stereochemical approach | 43
2.2 Envisioning new possibilities of chemical synthesis for biology and medicine | 57
2.3 Discovering the stereochemical mechanism of fermentation | 62
2.4 The heuristic role of the lock-and-key analogy in Fischer's program | 70

3 The making of the lock-and-key model of the antibody-antigen relationship, 1886-1930 | 77
3.1 Paul Ehrlich's understanding of immunological specificity and the lock-and-key analogy | 79
3.2 Origins of Ehrlich's receptor model | 86
3.3 The construction of the receptor model in the realm of immunology | 93
3.4 Receptor model reconstruction in terms of the lock-and-key analogy | 109

4 Lock-and-key foundations for molecular biology: Linus Pauling and the Caltech group, 1930-1960 | 133
4.1 Specificity: Immunochemical trends and traditions | 136
4.2 A new stereochemical view of antibody-antigen complementarity | 140
4.3 Pauling's universal molecular agenda: The importance of complementarity for biochemical reactions | 146
4.4 The lock-and-key analogy in science administration and cross-disciplinary communication at Caltech | 150
4.5 Postponing a paradigm shift at Caltech? | 168

5 Lock-and-key-based modeling and its influence on the development of biochemical research programs | 173
5.1 Roles of the analogy: From lock-and-key heuristics to lock-and-key reconstruction | 177
5.2 Analogical model reconstruction | 190

6 Concluding remarks on the construction of analogy-based research programs | 199

Literature | 205
Archival sources | 205
Primary sources (published) | 210
Secondary sources | 216

Preface

This book has emerged from a PhD project in the philosophy and history of science at the University of Bielefeld. When I started my PhD, I wanted to conduct a study on model transfer and epistemic interrelations between chemistry and the life sciences. In the course of the project, however, my understanding of the object of study and its historical development became much more pluralistic and context-sensitive. Especially the influence that research organization and historical reconstruction have on scientific modeling has transformed the initial project idea immensely. This combination of historical and philosophical perspectives is not an unproblematic one. Diverging methodologies, styles of reasoning, arguments and goals of study have to be accommodated. Although this process was quite a challenge, especially for the philosophical study of models and their role in science, it showed that historically oriented analyses have the potential to unveil essential contexts of scientific reasoning and development. I am grateful that I had the opportunity to conduct my research in an environment that supported me to use such epistemological tensions productively.

I would especially like to thank my advisors, Maria Kronfeldner and Carsten Reinhardt, for their open-mindedness and for all their guidance and valuable feedback throughout this process. I also owe special thanks to Martin Carrier and the philosophy department at the University of Bielefeld for significantly supporting my project and for providing such a productive research environment. I further wish to warmly thank Gregor Lax, Claudia Göbel, David Rengeling, Daniel Brooks, Stephan Kopsieker and Fabian Lausen for reading and discussing many ideas and drafts of this book. I would also like to thank Kenneth C. Waters, Mary Joe Nye, Marie Kaiser, Ulrich Krohs, William Bausman, Jack Powers, Martha Halina, Mads Goddiksen and Dana Mahr for discussing my work and their constructive suggestions.

Furthermore, this book could not have been written without the support of the Rockefeller Archive Center and the Oregon State University's Special Collections & Archive Research Center as well the Archive Center of the California Institute of Technology. I am especially grateful to Lee Hiltzik and Chris Petersen for their great help and for sharing their expertise on the collections. Nor would I have been able to carry out my research without the funding and support of the DAAD and the Bielefeld Young Researchers Fund.

Finally, I would like to say thank you to my family and friends, particularly Gregor Lax and Claudia Göbel, for supporting me intellectually and emotionally in countless ways that allowed me to write and finish this book. I cannot emphasize enough how grateful I am for your companionship.

Rebecca Mertens, Bielefeld,
September 2018

1 The lock-and-key analogy and its influence on 20th century biochemistry

The present study will focus on the "molecularization" of biochemistry – i.e. the establishment of the view that biochemical phenomena can be understood and controlled by investigating the structure of macromolecules and their functions in various kinds of biological processes.[1] This view became a cornerstone for the development of molecular biology and molecular medicine in the 1950s and 60s. Yet, despite the co-dependencies of the process of "molecularization" and the foundation of molecular biology in the second half of the 20th century, one should refrain from reducing the emergence of molecular biology to the successful "molecularization" of certain biological domains, as e.g. biochemistry, neurobiology and genetics.[2] Rather, there are certain events that put the emergence of molecular biology in a unique space, for instance a novel focus on nucleic acids as carriers and determinants of genetic material, the rise of a new metaphoric terminology within the "discourse of information",[3] as well as the devel-

---

1   Morange (1998): A History of Molecular Biology, Cambridge (Ma), p. 180; Kay (2000): Who wrote the book of life? A History of the Genetic Code, Stanford, p. 45.
2   Morange (1998): A History, p. 1f.; Rheinberger (1995): Kurze Geschichte der Molekularbiologie, in: Preprints of the Max Planck Institute for the History of Science (24), p. 2; Kay (1993): The Molecular Vision of Life, Oxford, p. 211.
3   Kay (2000): Who wrote the Book of Life?, pp. 73-127.

opment of experimental systems and social resources that by the 60s allowed interventions into the process of protein synthesis.[4]

However, some of the episodes, research techniques and concepts of macromolecular research in the early twentieth century are known to have had a considerable impact on the constitution of molecular biology. Especially the physico-chemical study of proteins, as well as the development of new physical instruments and modeling techniques that facilitated the chemical analysis of macromolecules in the early 20th century, such as ultracentrifugation, electrophoresis, and x-ray crystallography, shaped the very idea of a macromolecule, its biological functions and systematic attempts of intervention.[5] Furthermore, the concept of specificity and its use in the field of immunology and protein biochemistry served as a conceptual basis for the combination of chemical and physiological means of understanding natural processes.[6]

Specificity can be called a "loose concept" in that it had a multiplicity of meanings in the late 19th and 20th century and due to its vagueness served as a connection point between different branches of the biochemical and bio-

---

4   Rheinberger (1995): Kurze Geschichte, pp. 5-8; Kay (1989): Molecular Biology and Pauling's Immunochemistry: A Neglected Dimension, in: History and Philosophy of the Life Sciences (11), p. 211.

5   See Rheinberger and Müller-Wille (2017): The Gene from Genetics to Postgenomics, London and Chicago, p. 60. For the role of ultracentrifugation and electrophoresis in the preceding period of molecular biology, see e.g. Morange (2010): What history tells us XX. Felix Haurowitz (1896-1987) – A difficult journey in the political and scientific upheavals of the 20th century, in: Journal of the Biosciences, 35 (1), pp. 17-20, here p. 17. For the interplay of x-ray cristallography and stereochemical modeling in the analysis of organic molecules in the first half of the 20th century, see James (2014): Modeling the scale of atoms and bonds: The origin of space-filling parameters, in: Klein/Reinhardt (eds.): Objects of chemical inquiry, Sagamore Beach, pp. 281-320, here p. 283.

6   For historical studies on the concept of specificity and its influence on molecular biology, see Kay (2000): Who wrote, p. 41ff., Morange (1998): A History, pp. 12ff. and 15ff, and Mazumdar (1995): Species and Specificity. An interpretation of the history of immunology, Cambridge, Chapter 3 and 9.

medical sciences.[7] The common link of these different understandings of specificity was the assumption that some organisms could selectively react to external influences, e.g. to viruses or bacteria.[8] Conceptions of how this reaction should be characterized, however, differed in crucial respects. Some scientists and physicians, like the group of Robert Koch and Paul Ehrlich, thought of it as a "rigid and complete one-to-one relationship, according to which a given organism, constant as to morphology and physiology, causes a given disease", while others like for instance Carl von Nägli, Hans Buchner, and Karl Landsteiner pointed to the gradual nature of specific processes in the living organism.[9] Despite these differences in interpretation, the concept of specificity was "omnipresent in the biology of the first half of the twentieth century."[10]

As Morange points out, in most contexts of the early 20th century the concept of specificity was interpreted in the light of chemical theories and methods.[11] At the same time, chemistry became more and more 'physicalized' in that major innovations in the chemical sciences were accompanied by the development of instruments and measuring methods, which, in turn, were often created on the basis of theories from the 'new physics'.[12] In the course of these physico-chemical influences, a wide range of biochemists

---

7 For the notion of "loose concepts" and their role in the history of immunology, see: Löwy (1990): The strength of loose concepts: The case of immunology, in: History of Science, 30 (90), pp. 371-396. Anne Marie Moulin and Barbara Harshav pointed to the vagueness and the multiplicity of meanings that were ascribed to the concept of specificity in the 20th century. See Moulin and Harshav (1988): Text and Context in Biology: In Pursuit of the Chimera, in: Poetics Today, 9 (1), pp. 145-161, here p. 149 and 157.
8 Mazumdar (1995): Species, p. 82.
9 Ibid., p. 80.
10 Morange (1998): A History, p. 13.
11 Ibid.
12 Here, I mainly refer to the influences of thermodynamics on organic chemistry in the late 19th century and quantum physics in the late 1920s and 30s. See Nye (1993): From chemical philosophy to theoretical chemistry: Dynamics of matter and dynamics of disciplines, 1800-1950, Berkeley/London, pp. 227-261; Nye (1999): Before big science: The pursuit of modern chemistry and physics, 1800-1940, Cambridge (Ma), pp. 88ff. and 95ff.

investigated questions concerning the structure of macromolecules, as the latter were known to play fundamental roles in biological processes. The study of macromolecules became an important aspect of biochemical research in the early 20th century, because it promised insights into the connection between chemical structure and biological function. One of the underlying assumptions was that the role of macromolecules in biological processes was determined by the complementary fit of these molecules; in other words, that molecules were selected to bind and thus cause biochemical reactions on the basis of their structural fitting relationship. The origins of this idea have been ascribed to the German chemist Emil Fischer and his introduction of the so-called lock-and-key analogy.[13] In the 1890s Fischer proposed that enzymes and substrates must have complementary structures in order to cause a fermentation reaction and compared the enzyme-substrate relationship with the relation between a lock and a key.[14] As will be shown in the course of this study, the idea that the complementary lock-and-key-like fit between macromolecules could account for almost any biological phenomenon provided a simple and easily deployable, yet vague picture of how knowledge about chemical structure could be used for the exploration and explanation of biological functions.

The lock-and-key analogy re-appeared in several biochemical and biomedical contexts, and soon dominated the study of biomolecules. Frieder W. Lichtenthaler speaks of "a dogma for explaining principal life processes" that has been "induced" by the usage of the lock-and-key analogy in the context of early 20th century biomedicine and embryology.[15] He mentions two scientists in particular who have contributed to the pervasiveness of the analogy, the physician and Nobel Prize laureate Paul Ehrlich and the embryologist Frank R. Lillie. Lichtenthaler claims that both have used the

---

13 Morange (1998): A history, p.13; Mazumdar (1995): Species, p. 190-196; Kay (1993): A molecular vision, p. 173f.; and Kay (2000): Who wrote the book of life, p. 43.

14 Fischer (1894): Einfluss der Configuration auf die Wirkung der Enzyme I, in: Berichte der Deutschen Chemischen Gesellschaft, Vol. 27 (1894), pp. 2.985-2993. This article was also published in: Fischer (1906): Gesammelte Werke: Untersuchungen über Kohlenhydrate und Fermente II, ed. Max Bergmann, Berlin/Heidelberg, pp. 836-844.

15 Lichtenthaler (1994): Hundert Jahre, p. 2364.

analogy in a very speculative way, which eventually led to "a rather free, uncontrolled profilation of the concept from chemistry into medicine and biology."[16] He points out that the analogy "provided successive generations of scientists with their mental picture of molecular recognition processes, and, thus has shaped to a marked degree the development not only of organic chemistry, but, by extension to basic life processes, that of biology and medicine as well."[17] Scott F. Gilbert and Jason P. Greenberg argue along similar lines. They state that the concept of stereocomplementarity, which dominated biomedicine especially in the second half of the 20th century, was introduced and made accessible to biological and medical research by the usage of the lock-and-key analogy.[18] The authors assert that it "is hard to imagine any phenomenon on the cellular or molecular level which is not governed by lock-and-key stereocomplementarity."[19]

This study will examine the influence of the lock-and-key analogy in the first half of the 20th century by reconstructing its use in three influential research programs: (1) in Emil Fischer's studies on sugar fermentation and enzymatic reactions in the late 19th century, (2) in immunology and immunochemistry from the late 19th century until the 1930s and specifically in the research of Paul Ehrlich and his contemporaries; and finally (3) in research on the physico-chemical basis of macro-molecular specificity at the California Institute of Technology (Caltech) conducted by Linus Pauling and his colleagues from the 1930s to the 1960s. The aim of this case study is to specify how the lock-and-key analogy was used in these research programs and how it contributed to the establishment and manifestation of the concept of molecular complementarity and to the "molecular revolution" in 20th century biochemistry.[20]

---

16 Ibid., p. 2371.
17 Ibid., p. 2364.
18 Gilbert/Greenberg (1984): Intellectual Traditions, p. 18.
19 Ibid.
20 Rheinberger (1995): Kurze Geschichte, p. 2.

1.1 STATE OF THE LITERATURE

Much of the historical work on the origins of the lock-and-key analogy in late 19th century organic- and biochemistry is conducted by the scientists of these very fields.[21] These studies serve as a valuable source for locating the fields in which the lock-and-key analogy was used and for reconstructing disciplinary thinking. However, one must take into account that they are usually written in the context of disciplinary canonization and thus, among other purposes, serve to establish historical continuity between different episodes of biochemical thinking. What is more, such disciplinary constructions of history often emphasize what in retrospect has become a success story while alternative research strategies and concepts are often disregarded once a dominant view has been established.[22] Thus, while disciplinary accounts of the history of biochemistry and biomedicine underline the pervasive significance of the lock-and-key analogy they do not specify the different usages of the analogy and the varying contexts of scientific practice in which it appeared. Furthermore, they do not provide an answer to the question of why the analogy was appealing for biochemists. In order to understand how it could gain such a broad influence on biochemical research and education under these conditions, and how this impact is to be characterized in socio-epistemological terms, I will turn to the history and philosophy of science.

Historians of science and medicine have conducted studies on the lock-and-key analogy often linked to the history of influential individual scien-

---

21 See e.g. Hudson (1941): Emil Fischer's Discovery of the Configuration of Glucose, in: Journal of chemical education, pp. 353-357; Hudson (1948): Historical Aspects of Emil Fischer's Fundamental Conventions for Writing Stereo-Formulas in a Plane, in: Advances in Carbohydrate Chemistry, 3, pp. 1-22; Gilbert and Greenberg (1984): Intellectual Traditions in The Life Sciences II. Stereocomplementarity, in: Perspectives in Biology and Medicine, 28 (1), pp. 18-34; Lichtenthaler (1994): Hundert Jahre Schlüssel Schloss Prinzip: Was führte Emil Fischer zu dieser Analogie?, in: Angewandte Chemie, 106, p. 2456-2467; Barnett and Lichtenthaler (2001): A history of research on yeast 3: Emil Fischer, Eduard Buchner and their contemporaries, 1880-1900, in: Yeast, 18, pp. 363-388.
22 See Butterfield (1968 {1931}): The Whig Interpretation of History, London.

tists. In particular, two of the protagonists who are commonly associated with the lock-and-key analogy, Paul Ehrlich and Linus Pauling, and their research strategies have received much attention. Ehrlich (1854-1915) was a German physician and immunologist. He is most known for his side-chain theory of immunity for which he was awarded the Nobel Prize for physiology or medicine (together with Ilya Mechnikov) in 1908 and for the development of "Salvarsarn", the first chemotherapeutic drug against syphilis.[23] In the last 20 years, there has been a growing stock of historical research on Ehrlich's person and on the social infrastructure of the "Ehrlich school", his conceptual and methodological strategies,[24] and on his usage of pictorial

---

23 Note that chemotherapy was at first developed as a cure for infectious diseases. It was not until the 1940s that chemotherapeutics were used for the treatment of certain types of cancer. See Bruce A. Chabner and Thomas G. Roberts, Jr. (2005): Chemotherapy and the war on cancer, pp. 65-72.

24 Axel Hüntelmann has written a biography about Ehrlich with a particular focus on the social and economic infrastructure of Ehrlich's research groups (Hüntelmann (2011): Paul Ehrlich. Leben, Forschung, Ökonomien, Netzwerke, Göttingen. See as well Hüntelmann (2010): Legend of science: External constructions by the extended family – the biography of Paul Ehrlich, in: InterDisciplines 2, pp. 13-36). Hüntelmann also recently published an article on Ehrlich's role in the arising pharmaceutical industry in the early 20th century (Hüntelmann (2013): Making Salvarsan. Experimental Therapy and the Development and Marketing of Salvarsan at the Interface between Science, Clinic, Industry and Public Health, in: Jean-Paul Gaudillière and Volker Hess (eds.): Ways of Regulating Drugs in the 19th and 20th Centuries, Basingstoke, pp. 43-65). For studies on Ehrlich's conceptual and experimental practice, see Cambrosio, Jacobi and Keating (1996): Ehrlich's "beautiful pictures" and the controversial beginnings of immunological imagery, in: Isis, 84 (4), 662-692; Ead.(2004): Intertextualité et archi-iconicité: le cas des représentations scientifiques de la réaction antigène-anticorps, in: Études de communication, 27, pp. 2-13; Travis (2008): Models for biological research: The theory and practice of Paul Ehrlich, in: History and Philosophy of the life sciences, 30, 79-98; Mazumdar (1995): Species and specificity, Chapter 5, 6, 10 and 11; and Silverstein (2002): Paul Ehrlich's receptor immunology: The magnificent obsession, San Diego. For studies on Ehrlich's connections to the German dyestuff industry, see e.g. Lenoir (1988): A magic bullet: Research for profit and growth of knowledge in Germany around 1900, in:

representations.[25] Ehrlich has been linked to the history of the lock-and-key analogy in the context of his immunological and chemotherapeutic research on receptors, which he characterized as foodstuffs that bind the cell and form an intermediate linkage point between the cell and the antibody.[26] Especially the drawings that Ehrlich had used for the representation of the relationship between antibody and antigen, and also the one between cells and receptors, have often been described as variants or successors of the lock-and-key analogy by Ehrlich's contemporaries as well as by historians of science and medicine.[27] In the existing literature it remains open, however, which exact role the lock-and-key analogy has played in Ehrlich's research.

Linus Pauling (1901-1994) is known to have "confirmed" Fischer's hypothesis on lock-and-key-stereocomplementarity with his template model of antibody-formation, which found broad application in immunology, serology, molecular biology, and embryology in the mid-20th century.[28] He was awarded the Nobel Prize in Chemistry in 1954 and also received the Peace Prize in 1963 for his political commitment against nuclear armament.

Minerva 26 (1), pp. 66-88, and Travis (1989): Science as receptor of technology: Paul Ehrlich and the Synthetic Dye Stuffs Industry, in: Science in Context, 3, pp. 383-408.

25 See Cambrosio et al. (1993): Ehrlich's "beautiful pictures" and the controversial beginnings of immunological imagery, in: Isis, 84 (4), pp. 662-692.

26 Cambrosio et al. (2004): Intertextualité et archi-iconicité, pp. 5ff.; Morange (1998): A history of molecular biology, p. 13; Mazumdar (1995): Species and specificity, pp. 195f., p. 229 and p. 236; Lenoir (1988): A magic bullet, p. 75; Silverstein (2002): Paul Ehrlich's receptor immunology: The magnificent obsession, p. 83; and Lichtenthaler (1994): 100 Years "Schlüssel-Schloss-Prinzip", p. 2371.

27 See Cambrosio et al. (1993): Ehrlich's "Beautiful pictures", p. 682f.; Cambrosio et al. (2004): Intertextualité et archi-iconicité: le cas des représentations scientifiques de la réaction antigène-anticorps, in: Études de communication, 27, pp. 2-13, here p. 5-7; Travis (2008): Models for biological research: The theory and practice of Paul Ehrlich, pp. 88ff.; Morange (2000): A history of molecular biology, p. 13; Mazumdar (2002): Species and specificity, pp. 195f., p. 229 and p. 236.

28 Morange (1998): A History, p. 15.

Especially Pauling's frequent usage of molecular models for problem solving, his interdisciplinary "style", and his role in the institutionalization of molecular biology have been picked up extensively by historians and philosophers of science.[29] Lily Kay analyzes Pauling's contribution to the formation of molecular biology and ascribes it to his leading role at the California Institute of Technology (Caltech) and his close ties to the Rockefeller Foundation.[30] She further points out that Pauling's program on the nature and function of antibodies at Caltech had a huge impact on the "new

---

29 Pauling's biography and his contributions to the life sciences are well captured by Thomas Hager (1995): Force of nature: The life of Linus Pauling, Michigan; and Id. (1998): Linus Pauling and the chemistry of life, Oxford. Mary Jo Nye has analyzed Pauling's modeling techniques, his usage of "paper tools", as well as his interdisciplinary style of thinking between physics, biology and chemistry (see Nye [2001]: Paper tools and molecular architecture in the chemistry of Linus Pauling, in: Klein [ed.]: Tools and modes of representation in the laboratory sciences, Boston, pp. 117-132; and Nye [2000]: Physical and biological modes of thought in the chemistry of Linus Pauling, in: Studies in the history and philosophy of science, 31, pp. 475-491). Furthermore, Nye has conducted a study on Pauling's style of textbook writing (Nye [2000]: From student to teacher: Linus Pauling and the reformulation of the principles of chemistry in the 1930s, in: Lundgren and Bensaude-Vincent [eds.]: Communicating chemistry: Textbooks and their audiences, 1789-1939, European Studies in Science History and the Arts, 3, pp. 397-414). Recently, Jeremiah James analyzed the stereochemical and physical influences on Pauling's scale-modeling techniques (James [2014]: Modeling the Scale of Atoms and Bonds: The Origins of Space-filling Parameters, in: Klein and Reinhardt [eds.]: Objects of Chemical Inquiry, Sagamore Beach, pp. 281-320). Apart from studies that aim to capture Pauling's epistemic strategies, there has been research on his activities in research management and his interactions with science administrators (see e.g. Kohler [1991]: Partners in science. Foundations and Natural Scientists, 1900-1945, Chicago; and Kay [1989]: Molecular Biology and Pauling's Immunochemistry: A Neglected Dimension, In: History and Philosophy of the Life Sciences, 11 [2], pp. 211-219; Id. [1993]: A molecular vision, and Id. [2000]: Who wrote the book of life).
30 Kay (1989): Molecular Biology and Pauling's Immunochemistry: A Neglected Dimension, in: History and Philosophy of the Life Sciences, 11 (2), pp. 211-219; and Id. (1993): A molecular vision.

biology" for which the natural science division's director of the Rockefeller Foundation, Warren Weaver, coined the term "molecular biology" in 1938.[31] Kay briefly mentions the appearance of Fischer's lock-and-key analogy in a prominent article that Pauling published with the physicist Max Delbrück on the "nature of the intermolecular forces operative in biological processes" in 1940.[32] She states that in this article Pauling and Delbrück "extended Emil Fischer's lock-and-key model beyond enzyme-substrate-relations",[33] but does not analyze the role of Fischer's model or of the lock-and-key analogy in Pauling's other writings or projects. Cambrosio et al. deal with Pauling's visual models and especially with the pictures that he used for his 1940 paper on the chemical basis of antibody formation. These drawings were created in collaboration with the artist Roger Hayward.[34] The authors link Pauling's and Hayward's drawings to Ehrlich's "beautiful pictures" and Fischer's lock-and-key analogy,[35] but just like Kay and Morange, they do not specifically analyze the role of the analogy in Pauling's immunological and molecular program. It thus remains an open question in the historical literature if and how the analogy contributed to Pauling's research practice.

In general, works that relate the successes of the lock-and-key analogy in different fields and periods in the history of biochemistry and biomedicine to each other are lacking. The appearance of lock-and-key models in different branches of biochemistry and biomedicine (such as in olfaction research, embryology, neurology, and immunology) has been noticed by historians,[36] but so far it has not been related to the influencing role of the

---

31 Kay (1993): A molecular vision, pp. 10-13.
32 Pauling and Delbrück (1940): The nature of the intermolecular forces operative in biological processes, in: Science, 92 (2378), pp. 77-79.
33 Kay (1993): A molecular vision, p. 173.
34 Pauling (1940): A theory of the structure and process of antibody formation, in: Journal of the American Chemical Society, 62 (10), pp. 2643-2657. Cambrosio, Jacobi, and Keating (2005) analyze Pauling's and Hayward's drawings (See Cambrosio et al. (2005): Arguing with images: Pauling's Theory of Antibody formation, in: Representations, 89 (1), pp. 94-130).
35 Cambrosio et al. (2005): Arguing with images, p. 108f.
36 Recently, Carsten Reinhardt has pointed to the influential role of stereochemical ideas on theories of olfaction in the second half of the 20th century. He notes that

analogy in the 20th century. Thus, the history of the lock-and-key analogy and associated models in biochemistry remains rather fragmentary.

In the philosophy of science literature the lock-and-key analogy has been touched upon erratically. Kenneth Schaffner briefly mentions the analogy in the context of his work on reduction and particularly in his study on the operon model in molecular genetics.[37] Anne-Sophie Barwich analyzes receptor models and modeling techniques in olfaction chemistry.[38] She states that a characteristic element of lock-and-key modeling in this field of research was the transferability of molecular models from other disciplinary contexts. Lacking empirical knowledge about the shape of odor receptors, the lock-and-key model was used to make theoretical assumptions based on knowledge about other molecular mechanisms, most prominently from the field of enzymology. In this function the lock-and-key model dominated ol-

---

the so-called "stereochemical theory of olfactory perception" was based on a "rather crude 'lock-and-key' schema." In his depiction of the origins of the stereochemical theory in this field, Reinhardt focuses on the chemist John Earnest Amoore who proposed his theory in the early 1950s and thereby "adhered to a tradition of a long chain of similar attempts in chemistry, in the life sciences, and medicine, among the most famous of which was Emil Fischer's lock-and-key model of enzymatic reaction, and Paul Ehrlich's magic bullets." (Reinhardt [2014]: The olfactory object. Towards a history of smell in the 20th century, in: Klein and Reinhardt [eds.]: Objects of Chemical Inquiry, Sagamore Beach, pp. 321-341, here p. 324; 332). Furthermore, lock-and-key models have been located in other branches of neurophysiology and in the field of embryology. Gordon M. Shepherd argues that lock-and-key models played an important role in what he calls the neuroscientific revolution in the mid- and late 1950s (Shepherd [2010]: Creating modern neuroscience: The Revolutionary 1950s, New York, here p. 41f.), and Emily Martin calls attention to the ubiquity of lock-and-key models in 20th century embryology (Martin [1991]: The egg and the sperm: How science has constructed a romance based on stereotypical male-female roles, in: Signs, 16 [3], pp. 485-501, here p. 496).

37 Schaffner (1993): Discovery and Explanation in Biology and Medicine, Chicago, p. 481.
38 Barwich (2013): Making Sense of Smell: Classification and Model Thinking in Olfaction Theory, Doctoral thesis, University of Exeter, p. 155.

faction chemistry throughout the 20th century.[39] The anthropologist Emily Martin also mentions the lock-and-key analogy in her study on male/female attributions to egg and sperm concepts in the history of embryology in the 1980s.[40] Martin argues that the usage of lock-and-key terminology in this context resulted in the suppression of alternative models, especially of those that ascribed an active role to the female egg. According to this view, embryologists adopted the view that molecules of the sperm substance played an active role in the fertilization process, while the female egg molecules had a rather passive role, just like a lock that needs to be activated by the male key.[41] Although experimental results had suggested that the egg could be regarded as an active agent in the fertilization process, lock-and-key terminology played an important part in rendering these new insights invisible, which could have led embryological research into another direction.[42]

The present study shows, in line with Martin's claim, that the lock-and-key analogy indeed supported a reductionist view of biochemical processes, suggesting that biological phenomena could be sufficiently explained by the application of chemical theories and methods.[43] Yet I also argue that concentrating on this repressive role of the analogy reveals just one aspect of its history in biochemistry. In some contexts, this reductionist view left an open space for conceptual exploration and allowed cooperation between chemists and biologists. Focusing on cases of lock-and-key analogy usage in the fields of enzymology (chapter 2), immunology (chapter 3), and molecular biology (chapter 4) in the first half of the 20th century, the present study shows that the analogy also served as an instrument for the inclusion of different biochemical fields. Furthermore, it sheds light on the conditions for the analogy becoming so successful such as to suppress other concepts and models in other biochemical and biomedical contexts in the 20th century.

---

39 Ibid., p. 169.
40 Martin (1991): The egg and the sperm, p. 496.
41 Ibid., pp. 496ff.
42 Ibid.
43 This becomes especially clear in the context of Linus Pauling's usage of the lock-and-key analogy for the extension of his claims on stereocomplementarity to genetics, embryology and immunochemistry (See the present study, chapter 4).

In sum, although it is commonly assumed that the lock-and-key analogy had a huge influence on biochemical thought and education in the 20th century, it has not been specified yet how this influence is to be characterized. This is in part due to the lack of a long-term analysis of the analogy in its various scientific contexts. From a historical perspective, the present study can be seen as a contribution to such a long-term analysis. It concentrates on the role of the lock-and-key analogy in the making of research programs and reveals new aspects of the mutual interactions between analogy usage, model making and the organization of research in terms of epistemic, social, and political activities. I will follow the development of lock-and-key analogy usage from the 1880s to the 1960s, as in this period the lock-and-key analogy took on a bigger role in the foundation and expansion of a cross-generational research program on the molecular basis of biochemical phenomena. The aim of the historical analysis is thus to specify the various usages of the lock-and-key analogy in different stages and contexts of this program and its contributions to the molecularization of biochemistry and biomedicine in the early and mid-20th century.

1.2 THEORETICAL APPROACH: THE PHILOSOPHICAL ANALYSIS OF ANALOGIES IN SCIENCE

I will analyze the influential role of the lock-and-key analogy on biochemical research on the basis of philosophical theories about the nature and role of analogies and analogical modeling. The philosophical accounts which have hitherto been developed provide an important basis to address the relationship and mutual interrelations between analogy usage and model making in science.

My understanding of what analogies are and how they can be used in science is based on the canonical work of Max Black and Mary Hesse, and on the more recent interpretation of that work by Daniela Bailer-Jones.[44] As

---

44 Black (1962): Models and Metaphors, Cornell; Hesse (1966): Models and Analogies in Science, Notre Dame. Bailer-Jones discussed the work of Black and Hesse in several articles and in her book on "scientific models in the philosophy of science" (See Bailer-Jones [2000]: Scientific Models as Metaphors, in: Hal-

Bailer-Jones points out, Black and Hesse both grasp analogies as semantic relations that can be used to make inferences from a familiar phenomenon in order to gain knowledge about a new phenomenon which needs to be investigated.[45] According to the authors, this feature of analogies has proven to be especially fruitful in scientific discovery processes and more precisely for scientific activities involved in concept formation and theory construction.[46] As opposed to other philosophical attempts towards the explanation of knowledge generation processes, Black and Hesse proposed an idea of scientific practice in which scientists are often confronted with situations of uncertainty. A common way to deal with these uncertainties is the use of analogies. Drawing analogies thus enables scientists to gain access to the phenomenon in question in situations in which the phenomenon is not fully understood or not even identified yet.[47] In such situations, it is useful to borrow knowledge and experience from domains to which one has better access, e.g. from other scientific fields or daily-life situations. As means of creativity, analogies are thus considered to be important for scientific knowledge generation.[48] According to Black's interactional view, the creative force of analogies results from an interaction between old and new meanings of an analogy and, more precisely, from the transfer of older meanings to new contexts which "forces the audience to consider the old and the new meaning together."[49]

---

lyn [ed.]: Metaphor and Analogy in the Sciences, Dordrecht, pp. 181-198;Bailer-Jones [2008]: Models, Metaphors, and Analogies, in: The Blackwell Guide to Philosophy of Science, Chapter 6; Bailer-Jones [2009]: Scientific Models in Philosophy of Science, Pittsburg, pp. 46-80 and 106-126).

45 Bailer-Jones (2008): Models, Metaphors, and Analogies, p. 111ff.
46 Ibid., p. 112.
47 Bailer-Jones (2009): Scientific Models in the Philosophy of Science, Pittsburgh, p. 106.
48 Ibid.
49 Black (1977): More about Metaphor, in: Ortony (ed.): Metaphor and Thought, Cambridge, p. 38.

1.2.1 Metaphors, analogies, and modeling

According to Black the "use of (theoretical) models resembles the use of metaphors in requiring analogical transfer of a vocabulary. Metaphor and model-making reveal new relationships; both are attempts to pour new content into old bottles."[50] However, citing Toulmin, Black mentions that a model, at least if it is a "good" one, will be more than a metaphor, as "it is the suggestiveness, and systematic deployability, that makes a good model something more than a simple metaphor."[51] He further illuminates the relationship of models and metaphors as follows: "Since the basic analogy or root metaphor ... normally arises out of common sense, a great deal of development and refinement of a set of categories is required if they are to prove adequate for a hypothesis of unlimited scope."[52] Models are then seen as the elaborations of root metaphors and basic analogies; the goal of this elaboration being, according to Black, the creation of a hypothesis that is generalizable to all possible phenomena.

Daniela Bailer-Jones provides a compelling analysis of Black's interactional view and uses Black's depiction of the process of model-building out of "basic analogies" and "root metaphors" as a basis for her characterization of "metaphorical models".[53] Other than Black, she makes clear that there are also kinds of models which "develop in a different, non-metaphorical manner."[54] She grasps "metaphorical models" as a category of models that comprises analogical models. The basic idea behind this concept of "metaphorical models" is that they are in some ways built on the grounds of metaphorical, non-literal expressions.[55]

In line with Black, Bailer-Jones sketches the process of metaphorical model building as a process of elaboration that begins with the introduction of a metaphor and, via analogical inference between the target and the source domain, ends with a well-defined model.[56] The starting point in

---

50 Black (1977): More about Metaphor, p. 238f.
51 Ibid., p. 239.
52 Ibid., p. 240.
53 Bailer-Jones (2000): Scientific Models as Metaphors, p. 181; p. 186.
54 Ibid., p. 191.
55 Ibid., p. 192ff.
56 Bailer-Jones (2008): Models, Metaphors, and Analogies, p. 113f.

drawing analogies between two domains then is to be found in the conviction or even the certainty that these domains are similar in some respects. In most cases sacrifices are made, in that the resulting model is to a certain extent speculative; it "is tentative and unconfirmed in parts."[57] However, due to the initially identified similarities between the original and the new domain, the scientists using the metaphorical model would be confident that the study of these yet unconfirmed aspects leads to further clarification and discoveries.[58] In this sense the model can be called "suggestive" and "systematically deployable"; it allows to focus on hypotheses confirmed in another domain and thereby provides a system in which the study of still unknown phenomena (in the new domain) can proceed. A model which is built on the grounds of a metaphor and elaborated by analogical inference has thus "the capacity to encourage further investigative and creative development."[59]

Hence both Black and Bailer-Jones see metaphors and analogies as linked sequences in an elaborative process of modeling; while Black ascribes this process to the construction of "theoretical models",[60] Bailer-Jones prefers to narrow it to the group of "metaphorical models".[61] Yet Bailer-Jones also points to functional differences between metaphors and analogies on the one hand and metaphorical models on the other. The main purpose of models would be to enable access to scientific phenomena, whereas metaphors and analogies mainly function by transferring expressions from one domain to another.[62] In that function, the latter can also be used in order to gain access to phenomena, but it is not a mandatory element of metaphors and analogies to do that. According to Bailer-Jones, models and metaphors share that they are both descriptions,[63] whereas

---

57 Bailer-Jones (2000): Scientific Models as Metaphors, p. 194.
58 Black (1962): Models and Metaphors, p. 239. See also Bailer-Jones (2000): Scientific Models as Metaphors, p. 195f.
59 Bailer-Jones (2000): Scientific Models as Metaphors, p. 196.
60 Black (1962): Models and Metaphors, p. 239.
61 Bailer-Jones (2000): Scientific Models as Metaphors, p. 181.
62 Bailer-Jones (2008): Models, Metaphors, and Analogies, p. 124.
63 Ibid.

analogies "can exist as formal relationships between phenomena or rather, between the theoretical treatment of phenomena."[64]

Despite their structural differences, metaphors, models, and analogies also share a couple of functions and are often used together. Analogy, for instance "deals with resemblances of attributes, relations or processes in different domains."[65] Uncovering resemblance relations between a source and a target domain is, in turn, a beneficial and sometimes crucial condition for metaphorical transference processes and for modeling phenomena.[66] Mary Morgan also deals with the question of how metaphors, analogies and some kinds of models operate together in knowledge generation processes. In contrast to Bailer-Jones, Morgan speaks of "analogical models" instead of "metaphorical models."[67] Similar to Black and Bailer-Jones, Morgan describes the process of analogical model building as one in which a metaphor provides the basis for analogies and these, in turn, are developed into analogical models.[68] While the initial metaphor, according to Morgan, "suggests much, but tells us little", analogical models are much more concrete than analogies and metaphors; they put more constraints on "the world of the scientist."[69] Thus, the suggestiveness of metaphors provides scientists with the "raw material from which to make substantial analogies" which then allow for analogical model building.[70] But analogies already constrain the model to some extent; developing an analogical model means to use these constraints "as a way to explore the implications of that analogy."[71] Following Marcel Boumans, Morgan also speaks of developing multidimensionality when it comes to the cognitive process of turning meta-

---

64 Ibid., p. 111. Bailer-Jones is not always clear in her definition of analogies as relationships, as some passages suggest that analogies only "point to" or "deal with" resemblance relations. See ibid., pp. 110 and 124.
65 Ibid.
66 Ibid.
67 Morgan (2012): The World in the Model: How Economists Work and Think, Cambridge, p. 172f.
68 Ibid., p. 174.
69 Ibid., p. 173f.
70 Ibid.
71 Ibid.

phors into analogical models.⁷² A metaphor is thus "something one-dimensional" – in order to arrive at an analogical model from the basis of a metaphor, "a scientist needs to develop its various possibilities or dimensions into a model", in other words, she needs to create a world out of the metaphor.⁷³

1.2.2 Analogical modeling: Contexts and practices

The role of analogies in science has been an important topic in the context of the debate on the nature of models and their functions in scientific practice.⁷⁴ Especially when it comes to the creative processes that are involved in scientific modeling, analogies have been described as "archetypes", or "ideal types" for model building.⁷⁵ Amongst others, Nancy Nersessian has studied analogy use as part of model building processes.⁷⁶ In the tradition of Black and Hesse, Nersessian claims that analogies are crucial cognitive instruments for scientific productivity and innovation. One of her larger projects is to explain how reasoning by analogy works from the perspective of the cognitive sciences. According to Nersessian, analogical reasoning can best be grasped as a "model-based" strategy, meaning that it is first and foremost a means of cognitive model making.⁷⁷ This claim challenges a

---

72 Ibid., p. 174. Here, Morgan is referring to Boumans (2005): How Economists Model the World to Numbers, London.

73 Ibid., p. 174.

74 Frigg and Hartmann (2012): Models in Science, in: The Stanford Encyclopedia of Philosophy: http://plato.stanford.edu/archives/fall2012/entries/models-science, 10/08/2015, 14:00.

75 See e.g. Black (1962): Models and Metaphors, pp. 219-243, and Morgan (2012): A World in the Model, p. 141.

76 Nersessian examines strategies of analogy-based modeling in several of her monographs and articles, such as, Nersessian et al. (1999): Model-based reasoning in scientific discovery, Dordrecht; Id. (1999): Model-based reasoning in conceptual change, in: Magnani, Nersessian and Thagard (eds.): Model-based reasoning in scientific discovery, New York; Id. (2002): Model-based reasoning: Science, Technology and Values, Dordrecht, and Id. (2008): Creating Scientific Concepts, Cambridge (Ma).

77 Nersessian (1999): Model-based reasoning in conceptual change, p. 20.

view which has dominated the philosophical treatment of analogies up to the late 20th century, namely that the major goal of analogical reasoning in science is the construction of arguments.[78] The logical empiricists, and most famously Rudolf Carnap, claimed that analogical arguments are less powerful than inductive arguments.[79] Nersessian argues that the advocates of this claim have failed in locating the role of analogies in science. Analogies play important roles for scientific reasoning, just not so much for argumentative reasoning. Rather "the heart of analogy is employing generic abstraction in the service of model construction, manipulation and evaluation."[80] Hence, if one wanted to understand the scientific role of analogies, one has to analyze strategies of "model-based reasoning."[81] Nersessian distinguishes three types of model-based reasoning: "Analogical reasoning", "visual reasoning" and "thought experimenting". Characteristic for all these types of reasoning is that they lead to a change of perspective(s), which is in turn crucial for scientific creativity and progress. Referring to Max Black's interactional view, she states that analogical modeling "might help us to notice what otherwise would be overlooked, to shift the relative emphasis attached to details - in short, to see new connections."[82]

Nersessian's studies, which make use of empirical knowledge from the cognitive sciences as well as from historical case studies, contribute significantly to the specification of cognitive practices and the role of analogies in scientific understanding and learning. However, what is missing in this line of research is the appreciation of the various contexts of scientific practice.[83] In addition to the characterization of the cognitive process of model-

---

78 Ibid.
79 Ibid.
80 Ibid.
81 Nersessian (2008): Creating Scientific Concepts, p. 12.
82 Ibid., pp. 180f.
83 More recently, Nersessian concentrates on social aspects of scientific problem solving. Particularly interesting in this context is her work with McLeod on interdisciplinary problem solving and identity formation. However, the authors stay in the realm of the cognitive sciences and social psychology and do not adopt a sociological perspective. See McLeod/Nersessian (2017): Interdisciplinary problem solving. Emerging models in Integrative Systems Biology, in: European Journal of Philosophy of Science (fc, in press). See also Osbeck/

ing by means of analogy, my study provides an answer to the question in which contexts of scientific practice and for which purposes modeling by means of an analogy becomes important. Here, I specify three of these contexts: (1) Epistemic problem solving, (2) the historical reception and retrospective canonization of scientific fields and models, and (3) research management. I propose these three categories as a first orientation for the localization of different contexts in which modeling by analogy influences the course of research programs. I am positive that there are more of these contexts which could be uncovered through other case studies. However, the point that I make by introducing these three categories is that the philosophical literature on analogical modeling has focused almost exclusively on the influencing role of analogies on models in the context of epistemic problem solving. For the case of the lock-and-key analogy, it can, however, be shown that the analogy's impact on modeling is relatively small in this context, if compared to its influence on retrospective communication and research management. Hence, if the goal is to find out more about the role of analogy usage in long-term modeling processes and research development, the consideration of these other contexts of scientific practice is vital.

Another emphasis of the case study presented here lies on the power of reconstructive activities in the making of research programs, and especially on the influential role of model reconstruction processes in this respect. As will be shown in the course of the study, there are models, which seem to be based on a particular analogy, but are in fact retrospectively reconstructed in terms of that analogy. It is important to note, though, that the analogy still plays a crucial role in these cases. In fact, I argue that – if used for such reconstruction processes – analogies can influence the course of a research program even stronger. This is due to the multiple contexts of scientific practice in which model reconstruction by analogy takes place. Other than model construction, which is usually taken to be an epistemic, inner-scientific activity, model reconstruction often exceeds the epistemic context of scientific practice and is also used for non-epistemic purposes.

Nersessian (2017): Epistemic Identities in Interdisciplinary Science, in: Perspectives on Science 25 (fc).

1.3 METHODOLOGY

I analyze lock-and-key analogy usage in the context of research programs. In the philosophical discussion, the term 'research program' goes back to Imre Lakatos, who used it to support his claim that scientific development is governed by rationalizable, methodological rules.[84]

According to Lakatos, every research program possesses a "hard core" and a set of "auxiliary hypotheses" that serve to protect this core.[85] Research programs can further be characterized by two different sets of methodological rules which orient the program towards a certain direction, the "negative" and the "positive heuristic".[86] While the negative heuristic "specifies the 'hard core' of the programme" and indicates which research paths are *not* to be followed, the positive heuristic on the other hand suggests how to proceed with research, consisting of "a partially articulated set of suggestions or hints on how to change, develop" and "modify, sophisticate, the refutable protective belt."[87]

What I take from Lakatos' conception of research programs is the idea of the importance of programmatic continuity in science; more specifically the idea that scientists aim at rescuing an established research program rather than provoking revolutions and thus use strategies which lead to that continuity.[88] Considering programmatic continuity, Lakatos' thoughts on research programs come very close to Kuhn's characterization of "normal science" as a preservation and enlargement project of knowledge, which is

---

84 Carrier (2002): Explaining Scientific Progress: Lakatos' Methodological Account of Kuhnian Patterns of Theory Change, in: Kampis et al. (eds.): Appraising Lakatos: Mathematics, Methodology, and the Man, Dordrecht, pp. 53-71, here p. 10.
85 Lakatos (1970): Falsification and the Methodology of Scientific Research Programmes, in: Lakatos/Musgrave (Eds.): Criticism and the Growth of Knowledge (1970), p. 191.
86 Ibid., p. 192.
87 Ibid., p. 193.
88 Lakatos (1970): The Methodology of Scientific Research Programmes, in: Worral/Currie (eds.): Philosophical Papers, Vol. 1, p. 47.

already accepted by a respective community.[89] In this respect, Lakatos and Kuhn mainly differ in their estimation of how long the periods of continuity last and in their normative claims concerning the value of continuity for research development. While Kuhn views discontinuity and revolutionary episodes as a characteristic element of scientific projects, Lakatos emphasizes the long-term nature of successful research programs and the possibility to make scientific progress within these programs. According to Lakatos, such programs can avoid revolutions due to their inner structure. He claims that objections, anomalies or counterexamples are already expected in the beginning of research programs and that the positive heuristic is the strategy that predicts (and maybe even produces) anomalies and deals with them.[90] He further states that the amount of sophistication of such a theoretical design lies in the formulation of rules that allow for an unproblematic replacement of the initial conditions for the research program. Hence, the more sophisticated a scientific program is conceptualized and designed in the beginning; the more successful it will be in the long run.[91] This is where my account of research programs substantially differs from Lakatos'. In my understanding, continuity is established in the course of a program and is not a result of careful theoretical examination in advance and the successful application of methodological rules. Here, I follow Alan Musgrave who criticizes Lakatos for his emphasis on the conception of positive heuristics as a constitutive and empirical-autonomous preliminary for research programs.[92] As Musgrave emphasizes, Lakatos uses his considerations regarding positive heuristics as a supporting claim for the autonomy of theory

---

89 Kuhn (1970 {1962}): The Structure of Scientific Revolutions, Chicago, Chapter 2, here p. 10. For a better understanding of Kuhn's conception of normal science and the similarities between Kuhn and Lakatos, see Carrier (2002): Explaining Scientific Progress: Lakatos' Methodological Account of Kuhnian Patterns of Theory Change.
90 Lakatos (1970): The Methodology, p. 49f.
91 Ibid., p. 50f.
92 Other philosophers followed Musgrave in his critique on Lakatos' conception of the positive heuristics. In the 1970s a critical discussion emerged concerning whether such a heuristic can really foresee and avoid anomalies of a research program. Philosophers who have participated in this debate are e.g. Elie Zahar (1973), John Worrall (1978), Alan Chalmers (1979), and Martin Carrier (1984).

over experimental facts.⁹³ "It is just not true", Musgrave states, "that refutations of any specific variant of a research programme can be produced and digested by a clearly spelled out heuristic."⁹⁴ Nevertheless, he gives credit to Lakatos' concept of positive heuristics in terms of a "plan for solving [...] mathematical problem[s]."⁹⁵ Hence, "heuristic hints [...] can be found, to a greater or lesser degree, in all research programs, and [...] their importance cannot be underestimated."⁹⁶ As useful as those "heuristic hints" are, Musgrave mentions, they do not usually develop independently from the course of the respective research program (as constitutive preliminary of this research program).⁹⁷ However, the autonomy of theory over experiment is just one of the critical aspects of Lakatos' conception of research programs. The crucial point which goes beyond Musgrave's critique is that continuity within a research program is not only established by the adjustment of hypotheses and models in reaction to empirical anomalies. Rather, continuity in the long run is also a product of the re-interpretation of scientific facts and discovery processes, which in turn is a crucial part of the inner-scientific construction of history and education. This aspect of scientific development has already been mentioned by Kuhn in "The Structure of Scientific Revolutions."⁹⁸ In the context of his thoughts on the tendency of scientists to render scientific revolutions "invisible", Kuhn emphasizes that

"scientists and laymen take much of their image of creative scientific activity from an authoritative source that systematically disguises – partly for important functional reasons – the existence and significance of scientific revolutions. [...] As the source of authority, I have in mind principally textbooks of science together with both the popularizations and the philosophical works modeled on them. All three of these categories [...] have one thing in common. They address themselves to an already ar-

---

93 Musgrave (1976): Method or Madness? Can the methodology of research programmes be rescued from epistemological anarchism, in: Id. (ed.): Essays in Memory of Imre Lakatos, Boston Studies in the Philosophy of Science, 39, pp. 457-491, here p. 468.
94 Musgrave (1976): Method or Madness, p. 470.
95 Ibid.
96 Ibid.
97 Ibid., p. 471.
98 Kuhn (1970 {1962}): The Structure of Scientific Revolutions, Chicago.

ticulated body of problems, data, and theory, most often to the particular set of paradigms to which the scientific community is committed at the time they are written. [...] The result is a persistent tendency to make the history of science look linear or cumulative, a tendency that even affects scientists looking back at their own research."[99]

The study presented here takes these re-interpretation processes into account and views them as crucial episodes in the establishment of programmatic continuity. I claim that the retrospective interpretation of scientific discoveries becomes especially important in transition phases in which the scope of a research program is expanded to other domains of research. As will be shown in the course of the study, analogies play an important role in these phases, as they can be used to create a linkage point between different research programs. This linkage, following the thesis, is provided by the re-interpretation and unification of models by means of analogy usage in the context of retrospective science communication.

I distinguish between two different types of research programs: Individual or single research programs which are, as the term implies, bound to the interests, questions and problems raised by individual scientists (e.g. "Fischer's stereochemical program on fermentation"), and long-term, cross-generational programs. It is important to note that individual programs can of course have many followers and that knowledge generation in these programs can certainly be seen as a social group activity. However, often these programs are associated with a founder, a scientist who has coined certain terms or who has proposed a theory or method that soon required canonical status within a scientific community. Cross-generational long-term programs, on the other hand, often have more than one founder; usually each generation ascribes groundbreaking changes to a prominent key figure (not uncommonly Nobel Laureates).[100]

For this study I looked at three individual research programs in which the lock-and-key analogy was used. Two of these research programs (Fischer's and Ehrlich's programs from the late 1880s to the early 20th century) were part of the same generation, but belong to different scientific

---

99 Ibid, pp. 136ff.
100 See also Kuhn (1970 {1962}): The Structure of Scientific Revolutions, p. 139f.

fields. The third program (Pauling and his co-workers at Caltech) started forty years later than the other two. There is a synchronic and a diachronic dimension of my analysis of the lock-and-key analogy. Looking at the synchronic dimension, I characterize the role of the analogy in the previously named individual programs.[101] The diachronic dimension, seen in terms of cross-generational change from Fischer through Ehrlich and Pauling to other biochemists at Caltech, allows us to see connections and differences in the previously analyzed individual research programs. Furthermore, it elucidates how scientists make references to previous programs and which role these references play in agenda setting and realization.

1.4 OUTLINE

The present study is divided into theoretical and empirical chapters. Chapter 2-4 represent cases of lock-and-key analogy usage in different, but sometimes related biochemical and biomedical research programs from the late 19th to the mid-20th century. Chapter 5 and 6 then aim to reflect and interpret the previous historical findings.

In Chapter 2, I analyze the origins of the lock-and-key analogy in enzymatic chemistry in the late 19th century and its role for the beginnings of the biochemical study of macromolecules. The focus is on the research of Emil Fischer who introduced the lock-and-key analogy in 1894 in the context of his studies on the stereochemical foundation of fermentation. I will show how Fischer used the analogy to apply the stereochemical approach that he had developed earlier in his studies on sugar classification to fermentation research. I will argue that this transfer of molecular geometry concepts from one domain of research (research on the structure of sugars) to another (research on the chemical relationship between enzymes and sugars during the process of fermentation) was facilitated by the usage of the lock-and-key analogy in a heuristic way. Specifically, I will show how

---

[101] One might object that this is also a diachronic perspective, as these programs of individuals or locally connected groups also undergo change. This is of course true. However, these changes are smaller and different with respect to analogy usage than the ones that become visible in the cross-generational dimension.

Fischer, with the help of the analogy, applied three heuristic strategies in his research practice: incompleteness, abstraction and simplification. I argue that in this way, a new research program on the stereochemical mechanism of fermentation was established. In addition, Fischer also drew on the analogy to lay out (and thus justify) the potential of his studies envisioning the future of biochemical research as the investigation of biochemical problems in terms of lock-and-key-like macromolecular fit.

Chapter 3 deals with the use of the lock-and-key analogy in immunology and chemotherapeutic research at the end of the 19th and the beginning of the 20th century. Focusing on the work of Paul Ehrlich, I will examine the role of the analogy for the application of macromolecular analysis in fields of biomedicine. Ehrlich is known for his application and introduction of the lock-and-key analogy to the field of immunology and chemotherapy.[102] However, I will argue that in effect it was the historical reconstruction of Ehrlich's work by his successors that subsumed Ehrlich's ideas under the lock-and-key model of antibody-antigen reactions. I will therefore examine the construction of Ehrlich's receptor model in toxicology and immunology, which was formulated on the basis of the side chain theory of the formation of specific antibodies. Throughout this work and the subsequent transfer of concepts to the study of chemotherapeutic agents for the treatment of infectious diseases, Ehrlich did not use lock-and-key terminology but applied his own toxicological heuristics and terms. I will show that from around 1900 until the 1930s, in several phases of the reception of Ehrlich's work by successors in the fields of immunology and medicine as well as in popularization contexts, the receptor model was re-constructed as the lock-and-key model of antibody-antigen reactions.

These developments will be taken up in Chapter 4, which investigates the role of the lock-and-key analogy in the founding periods of immuno-

---

102 See Cambrosio et al. (1996): Ehrlich's "Beautiful pictures", p. 682f.; Cambrosio et al. (2004): Intertextualité et archi-iconicité, p. 5-7; Travis (2008): Models for biological research, pp. 88ff.; Morange (1998): A history, p. 13; Mazumdar (2002): Species, pp. 195f., p. 229 and p. 236; Lenoir (1988): A magic bullet: Research for profit and growth of knowledge in Germany around 1900, in: Minerva 26 (1), pp. 66-88, here p. 75; Silverstein (2002): Paul Ehrlich's receptor immunology: The magnificent obsession, p. 83; and Lichtenthaler (1994): 100 Years "Schlüssel-Schloss-Prinzip", p. 2371.

1. Influence of the lock-and-key analogy on 20th century biochemistry | 35

chemistry and molecular biology from the 1920s to the 1960s in the United States. In the contexts under consideration, lock-and-key-like relations were generalized for the study of biologically important processes, such as fertilization, embryological development and genetic inheritance. At first, I will take a look at how immunochemists in the 1920s and 30s referred to the lock-and-key analogy in order to establish a continuum between Ehrlich's model of immunological processes and new theories of antibody formation, such as the lattice-, or framework theories. On this basis, the chemist Linus Pauling formulated the template model of antibody formation, relying on the idea of protein folding, which emerged as one of the central dogmas in molecular biology before the conceptual shift to the genetic code in the mid-1950s.[103] Pauling's template model was built on the idea of a lock-and-key-like fit between antibody and antigen. I will take a look at how this model explained the process of antibody formation as a reconfiguration that was needed to create a fit with the antigen. Furthermore, it will be shown that Pauling extrapolated from his antibody research and used the template model to ask how proteins in general are spatially transformed in order to perform different functions in the living organism. I will examine how Pauling communicated this generalized vision for the emerging cross-disciplinary field of molecular biology and how specialists from contributing disciplines, such as embryology, genetics and cell biology, took up his physico-chemical program around the study of proteins. The crucial point here is that this process of appropriation, and interdisciplinary cooperation was once more facilitated by using the lock-and-key analogy. The latter provided a fertile link between various research agendas creating major interdisciplinary research projects in the 1940s and 50s. For this context, I will analyze the use of the analogy in two exemplary cases in order to locate the role of the analogy in joint research proposals as well as on the level of research practice by the Caltech group.

After dealing with the role of the lock-and-key analogy case by case, chapters 5 and 6 will take an overarching perspective on what happened along the way from Fischer's introduction of the lock-and-key analogy in the 1890s to its usage in the context of Caltech's specificity program in the 1940s and 50s. Chapter 5 will summarize and interpret the findings from

---

103 Kay (2000): Who wrote the book of life, p. 51ff.; Morange (1998): A History, p. 15.

the case studies on the use of the analogy in the hitherto considered biochemical contexts in the first half of the 20th century. The main thesis that will be specified and defended in this chapter is that the lock-and-key analogy helped to establish a long-term, cross-disciplinary research program on the physico-chemical basis of macromolecular interactions. This long-term program, I argue, was a result of linking the individual research programs of Fischer, Ehrlich, and Pauling to each other by means of lock-and-key analogy usage. Chapter 6 will then elaborate on the idea that analogies can be "anchors" in the expansion of research programs.[104] Here, I will argue that analogies differ from other non-metaphorical anchors (e.g. mathematical idealizations) in that they play a crucial role in the reception and in the communication of scientific achievements to non-scientific and cross-disciplinary scientific audiences. This role of the analogy in different contexts of science (including science popularization and administration), I claim, makes a substantial difference when it comes to the expansion of a research program and to the impact of that program on other scientists and fields.

1.5 SOURCES

I have looked at both primary and secondary sources in order to specify the role of the lock-and-key analogy in the considered programs. The primary sources included published material, such as research articles, speeches, scientific monographs, popularizing science books (fictional and non-fictional), as well as archival material, i.e. laboratory notebooks, letters, research proposals and reports.

Chapters 3 and 4 are in part based on archival material, whereas chapter 2 relies on Fischer's published work and more specifically on his articles on sugar synthesis and fermentation, most of which were published in the *Reports of the German Chemical Society (Berichte der Deutschen Chemischen Gesellschaft)* between 1888 and 1907. For my analysis of Ehrlich's

---

[104] Christopher Pincock introduced this terminology, but has only used it for the class of mathematical idealizations and models. See Pincock (2012): Mathematical models of biological patterns: Lessons from Hamilton's selfish herd, in: Biology & Philosophy, 27(4), pp. 481-496.

1. Influence of the lock-and-key analogy on 20th century biochemistry | 37

immunological and chemotherapeutic program (chapter 3) and the role of the lock-and-key analogy in Linus Pauling's research program at Caltech (chapter 4), I visited several archives including the Rockefeller Archive Center (RAC), the Special Collections and Archive Research Center of the Oregon State University (OSU), and the Archives of the California Institute of Technology (Caltech Archives). In order to conduct the analysis of lock-and-key analogy usage in Ehrlich's program, I made use of the *Ehrlich papers* (RAC), in particular of *Ehrlich's notebooks* and the *Ehrlich Blöcke*, a collection of 1500 cards including notes that Ehrlich had written for himself, his secretary or for his assistants. These notes allowed insights into Ehrlich's style of reasoning and administration; they were especially informative with respect to the transfer of ideas and the role of theory-laden terminology in the communication of experimental results and orders. Another source which became important in the study of Ehrlich's reception and the re-interpretation of Ehrlich's ideas in terms of the lock-and-key analogy was the collection of newspaper clippings and articles at the RAC. This collection included articles from weekly local newspapers (e.g. *Allgemeine Frankfurter Zeitung, Die Frankfurter Nachrichten, Die Illustrierte Frankfurter Woche, Die Wiener Freie Presse*) as well as journal articles, mostly from medical journals (e.g. *Die Deutsche Medizinische Wochenschau, Wiener klinische Wochenschrift,* and *Die Naturwissenschaften)*.

Moreover, I made use of the *Ava Helen and Linus Pauling Papers* at the OSU Archives in order to specify the role of the lock-and-key analogy in Pauling's interactions with other scientists, in his efforts of science popularization and interdisciplinary education, as well as in his creative processes (chapter 4). In order to get insight into Pauling's administrative activities at Caltech, and his close connections to the Rockefeller Foundation, I looked at files on the *Division of Chemistry and Biology at Caltech* as well as on *Pauling's research on immunology,* all of which were located at the RAC. In particular, I have looked through the correspondence between Pauling and Warren Weaver, the director of the Natural Science Department of the Rockefeller Foundation (RF) from 1932 to 1955, at the *Warren Weaver Diaries,* as well as grant proposals and reports from Caltech to the RF. Also, the collection of the Caltech Archives related to the administrative and scientific work conducted by individual scientists and groups in Caltech's chemistry and biology departments was of special importance for chapter 4.

2 The lock-and-key analogy in Emil Fischer's program on sugar fermentation, 1890-1907

The lock-and-key analogy was first used in the context of enzyme and fermentation chemistry by the German organic chemist and Nobel laureate Emil Fischer.[1] In his first paper on enzymatic reactions in 1894,[2] Fischer speculated about the chemical mechanism of sugar fermentation and stated that "enzyme and glucoside must fit each other like lock and key in order to have any chemical effect on each other."[3] This phrase has since been cited many times by chemists and biochemists as well as by historians,[4] and has been referred to as a central "dogma for explaining principal life processes" in the first half of the 20th century.[5]

---

1 Mazumdar (1995): Species, p. 195.
2 Fischer (1894): Einfluss der Configuration auf die Wirkung der Enzyme I, in: Berichte der Deutschen Chemischen Gesellschaft, Vol. 27 (1894), pp. 2.985-2993. This article was also published in: Fischer (1906): Gesammelte Werke: Untersuchung über Kohlenhydrate und Fermente II, ed. Max Bergmann, Berlin/Heidelberg, pp. 836-844.
3 Fischer (1894): Einfluss der Configuration I, in: Berichte, p.2992; translated by Mazumdar (1995): Species, p. 198.
4 Barnett and Lichtenthaler (2001): A history of research on yeast 3: Emil Fischer, Eduard Buchner and their contemporaries, 1880-1900, in: Yeast, Vol. 18, pp. 363-388, here p. 377.
5 Lichtenthaler (1994): Hundert Jahre, p. 2364. See as well Gilbert/Greenberg (1984): Intellectual Traditions, p. 18.

As Mazumdar notes, Fischer's usage of the lock-and-key analogy has to be seen in the light of the increasing popularity of theories about the spatial dimension of molecules in late 19th century German organic chemistry.[6] In the period between 1890 and 1898, Fischer repeatedly addressed the question of how the 'geometry' of organic molecules influenced their chemical and biological behavior, such as their optical activity and their fermentation.[7] This question became especially important in the context of his work on sugar classification and synthesis.[8] In a talk on the importance of carbohydrate chemistry for the physiological and biological sciences, Fischer summarized his findings and concluded that

"[w]e have before us a quite new and I might say astonishing fact, that the most common function of a living being [fermentation, R.M.] depends more on the molecular geometry than on the composition of its food material."[9]

By "molecular geometry" Fischer referred to the three-dimensional structure of organic molecules.[10] As he mentioned himself, the question of how

---

6 Mazumdar (1995): Species, p. 182 and p. 195.
7 Passages in which Fischer addressed the issue of molecular geometry can be found in: Fischer (1890): Synthesen in der Zuckergruppe, in: Id. (2006) Untersuchungen über Kohlenhydrate, p. 28; Fischer (1894): Synthesen in der Zuckergruppe II, p. 54f., in: Untersuchungen; Fischer (1894): Die Chemie d. Kohlenhydrate und ihre Bedeutung für die Physiologie, in: Untersuchungen, p. 112; Fischer (1894): Einfluss der Configuration auf die Wirkung der Enzyme, p. 843; Fischer (1895): Einfluss der Configuration auf die Wirkung der Enzyme III, p. 859; Fischer (1898): Die Bedeutung der Stereochemie für die Physiologie, in: Untersuchungen, p. 134; Fischer (1899): Ueber die Spaltung racemischer Verbindungen in die activen Componenten, Berichte (32), p. 3; Fischer (1906): Organische Synthese und Biologie (Faraday Lecture), in: Untersuchungen auf verschiedenen Gebieten, p. 67.
8 Mazumdar (1995): Species, p. 193.
9 Fischer (1894): Die Chemie der Kohlenhydrate und ihre Bedeutung für die Physiologie. Rede gehalten zur Feier des Stiftungstages der Militärärztlichen Bildungsanstalten am 2. August 1894, in: Id. (1906): Untersuchungen über Kohlenhydrate, pp. 96-115, here p. 108. The original passage was translated by Mazumdar (1995): Species, p. 194.

2. The lock-and-key analogy in Fischer's research on fermentation | 41

the spatial arrangement of organic molecules determined their role in the chemical processes within living organisms was a "task of the greatest importance for biology."[11]

The idea that the three-dimensional structure of molecules had an effect on the biological functions of these molecules was not completely new: There were in fact several predecessors to Fischer who formulated molecular theories of biological phenomena, particularly Louis Pasteur in the 1860s[12] and, among other German dye stuff chemists, Otto Nikolaus Witt who started to develop his molecular theory of dyes in 1876.[13]

Yet, as will be shown in the course of this chapter, Fischer's attempt at studying biochemical phenomena contributed something new to previous approaches due to new systematic devices of molecular modeling (i.e. the so-called Fischer-conventions) and due to a more fine-grained understanding of the concept of molecular geometry residing therein. The analysis of the three-dimensional structure of organic molecules by means of paper tools and material modeling remained at first in the realm of organic chemistry, and was then, in the course of the 1930s and 40s, transported into biochemical and biomedical fields.[14] As the present study will show, the lock-and-key analogy and its initial scientific context, the study of sugar fermentation, provided an icon for the systematic exploration of molecular geometry for non-chemists, i.e. for biologists, physicians and physicists, in the early and mid-20th century. Yet, before I will come to these other scientific contexts, I will first turn to Fischer's research on sugar fermentation.

In his work on the sugar group, Fischer created a method to analyze the spatial arrangements of atoms in sugar molecules before the space structure

---

10 Fischer (1894): Die Chemie der Kohlenhydrate, p. 104f.
11 Fischer (1894): Einfluss der Configuration auf die Wirkung der Enzyme, in: Gesammelte Werke: Untersuchung über Kohlenhydrate und Fermente II, ed. Max Bergmann (Berlin/Heidelberg: Springer, 1906), p. 108. The translation of the original passage is taken from Mazumdar (1995): Species and Specificity: An Interpretation of the History of Immunology (Cambridge, MA.: Havard University Press), p. 194.
12 See Gilbert and Greenberg (1984): Intellectual Traditions, p. 19-22.
13 Travis (1991): Paul Ehrlich: 100 years of chemotherapy, 1891-1991, in: The Biochemist, Vol. 13, p. 10.
14 Morange (1998): A History, p. 14.

of molecules was fully empirically accessible and physically proven.[15] Up to the 1890s, conceptions of molecular geometry provided a useful tool for the explanation and prediction of chemical isomers, but were perceived as mere theoretical constructs.[16] As a powerful rhetorical and epistemic tool, the concept of molecular geometry played a central role in Fischer's research program on sugar classification and synthesis. I will show that this was a first step in the attempt to present the practice of organic synthesis within the new establishing framework of stereochemistry[17] as a systematic and thereby reliable alternative to previous chemical and biological means of intervention. As will be demonstrated, the second step was the transference of Fischer's stereochemical claims to the study of enzymes and sugar fermentation by means of the lock-and-key analogy. The hypothesis that molecular geometry was a decisive factor when it came to the distinctiveness of the sugars provided a first orientation in his attempt to explain the mechanism of enzymatic sugar fermentation.

---

15 Until the 1950s, it was only possible to determine relative molecular configurations. Relative configuration formulas depicted the relationship between the (then unknown) absolute configurations. The crystallographer Johannes Martin Bijovet was the first to determine the absolute configuration of a molecular compound (sodium-rubidium-tartrat - a salt of tartaric acid) by means of X-ray diffraction methods in 1951. See: André Authier (2013): Early Days of X-ray Crystallography, Oxford.

16 Ramberg (2003): Chemical Structure, Spatial Arrangement. The early history of stereochemistry, 1874-1914, (Burlington: Ashgate), p. 325.

17 Stereochemistry aims to determine the spatial arrangement of atoms within organic molecules and deals with the properties of stereoisomers, substances that have equivalent chemical structures if considered in a two dimensional space, but are different with respect to their three-dimensional arrangement. See Hargittai et al. (2008): Symmetry through the eyes of a chemist, Heidelberg, p. 97f.

2.1 ORIGINS OF THE CONCEPT OF MOLECULAR GEOMETRY AND FISCHER'S STEREOCHEMICAL APPROACH

By the mid-19th century, organic chemists had started to develop ideas of molecular geometry to explain the issue of optical isomerism. Pasteur and Mitcherlich were the first to observe that tartaric acid and racemic acid were optical isomers, meaning that they reacted differently against polarized light,[18] although they were known to be equal in their chemical structure. In order to deal with this kind of optical isomerism, Pasteur assumed that isomeric substances differed with respect to their molecular symmetry or asymmetry. By this he referred to the geometrical form of crystallized organic compounds. In his experiments with the salt derivatives of tartaric acid, he found that these salts (in crystallized form) consisted of two isomorphic, but antipodal groups.[19] They were isomorphic with respect to hitherto known criteria of chemical structure, but Pasteur supposed that their geometrical forms were asymmetrical; he imagined the two groups to be arranged in two reversed tetrahedra which were related to each other like mirror images ("Bild und Spiegelbild"). He concluded that only such an asymmetric molecular constitution would cause a substance to be optically

---

18 In the mid-19th century, chemists started to employ measurement methods from the realm of optics to determine the optical rotary power of organic substances, the way in which these substances affect the direction of polarized light. Polarimetry gained practical importance in the second half of the 19th century, especially in the industrial production of sugars and other foodstuffs. See Mund (1999): Struktur, Konfiguration, p. 123.

19 Pasteur (1860): Ueber die Asymmetrie bei natürlich vorkommenden organischen Verbindungen. 2 Vorträge gehalten am 20. Januar und 3. Februar 1860 in der Société chimique zu Paris, translated and edited by Ladenburg, Leipzig, p. 13. For a more detailed analysis of Pasteur's approach, see also: Mund (1999): Struktur, Konfiguration und Formelschreibweise der Kohlenhydrate von Kekulés und Coupers Valenzlehre (1858) bis zum Beginn von Emil Fischers Arbeiten über die Kohlenhydrate (1890), Hamburg/Bukarest, p. 113ff.

active and that this would only apply to substances occurring in the living organism, and not to artificial ones and minerals.[20]

Almost 30 years later, Johannes Wislicenus raised the issue of isomerism again and called for a theoretical solution.[21] In the meantime, organic chemists discovered that cases of optical isomerism were not exceptional, but rather an essential outcome of organic synthesis, the artificial production of organic substances.[22] However, the differences between isomers, though well known, could still not be represented by chemical formulas. This led Wislicenus to announce the isomerism issue as a deficiency of the traditional formula system and to call for a general transformation of the same.[23] Jacobus Henricus van't Hoff, a student of August Kekulé and Charles Adolphe Wurtz,[24] approached this task in his dissertation and published his first considerations of the three-dimensional structure of organic molecules in 1875.[25] His work was translated into German in 1877 and later

---

20 Gilbert and Greenberg (1984): Intellectual Traditions, p. 20; Mund (1999): Struktur der Kohlenhydrate, p. 113.
21 Wislicenus (1888): Ueber die räumliche Anordnung der Atome in organischen Molekülen und ihre Bestimmung in geometrisch ungesättigten Verbindungen (in zwei Bänden), in: Akademie der Wissenschaften. Abhandlungen (Math. Physik. Klasse) Band 14, Leipzig, p. 6.
22 See Mund (1999): Struktur, Konfiguration, p. 113.
23 Ibid., p. 30. See as well Rocke (2010): Image and Reality. Kekulé, Kopp, and the Scientific Imagination, Chicago, p. 239.
24 Both Kekulé and Wurtz became known for their substantial contributions to the developing field of organic chemistry in the mid-19th century. Charles Adolphe Wurtz (1817-1884), a French chemist and physician, is especially known for developing new methods for organic synthesis, and particularly for the synthesis of alkanes. August Kekulè (1829-1896) is one of the founders of structural organic chemistry in Germany and well known for his valence theory, the introduction of new forms of representation for the depiction of chemical structure, and in particular for his representation of the structure of benzene. See Brock (1997): Viewegs Geschichte der Chemie, Braunschweig/Wiesbaden, p. 147f. and pp. 156-163.
25 Van't Hoff (1875): La chimie dans l'espace, Rotterdam.

2. The lock-and-key analogy in Fischer's research on fermentation | 45

referred to as a cornerstone in the field of stereochemistry.[26] Van't Hoff contributed to the nascent stereochemical movement by specifying the asymmetry of a molecule and thereby proposing a concrete model for the arrangement of atoms in a three-dimensional space. Other than Pasteur's student, Joseph-Achille Le Bel, who also developed a theory of molecular

Figure 1

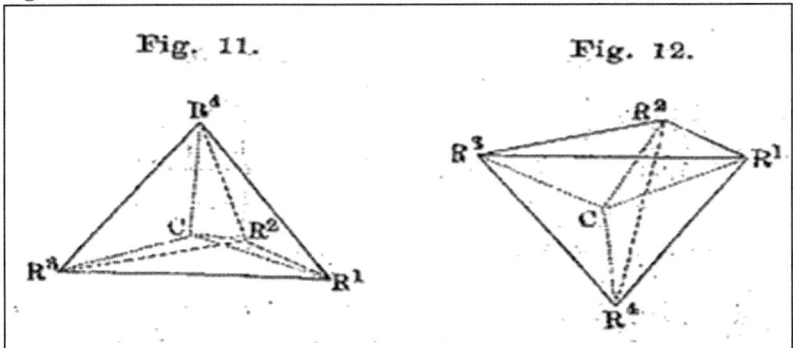

Figure 2

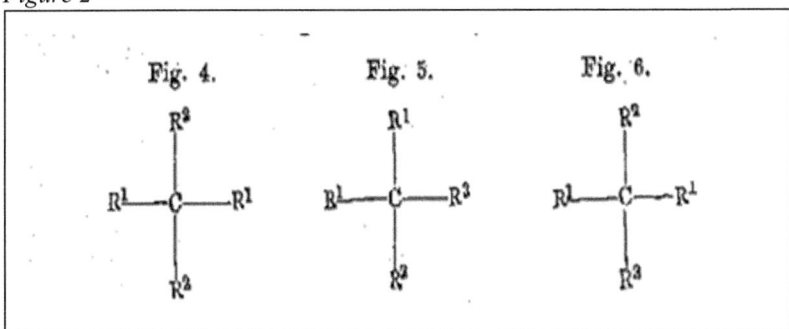

Figure 1 and 2: Van't Hoff used the above figures to explain his theory of asymmetric carbon atoms.[27] Both figures show the arrangement of the substituents around the carbon atom. Figure 1 depicts these arrangements within a tetrahedral space structure, while figure 2 shows the same from a two-dimensional perspective.

---

26 Ramsay (1975): Van't Hoff-Le Bel Centennial, ACS Symposium Series (Washington, DC), p. 74.
27 Van't Hoff: Lagerung der Atome im Raume, Braunschweig, p. 46.

space structure and published it in 1874, van't Hoff didn't approach the question of how organic molecules can be arranged in a purely mathematical manner.[28] Instead, he made use of material card board models in order to visualize the different possibilities of how the atoms could be spatially arranged within organic molecules.[29]

Van't Hoff's theory can be divided into two major hypotheses: (1) The assumption that isomers would only occur if the molecule consisted of at least one asymmetric carbon atom, i.e. a carbon atom that is directly linked to four or more different substituents (e.g. hydrogen, oxygen, sulfur, and nitrogen atoms), and (2) that valid conclusions about how many different substituents are linked to the carbon atom can only be made when the molecule is imagined as a three-dimensional tetrahedron with the carbon atom at the center; a two-dimensional representation of the molecular structure would not allow us to see all possible arrangements of the substituents.[30] These hypotheses allowed van't Hoff to predict the number of isomers for substances containing n asymmetric carbons, which increases for each substance according to 2n. That is, one asymmetric carbon atom would allow two isomers, two asymmetric carbon atoms would allow four isomers, three would allow eight isomers and so on.[31]

2.1.1 Theory as a tool for systematization in Fischer's sugar program

Van't Hoff's theory of the asymmetric carbon atom strongly influenced the development of carbohydrate chemistry in the late 19th century and gained special importance for the industrial production of sugars during the 1880s

---

28 Although both Le Bel and van't Hoff used mathematical reasoning, their approaches differed substantially from each other. While Le Bel searched for general principles for the emergence of optical activity, Van't Hoff concentrated on the properties of the carbon atom and formulated his theory on the grounds of structural organic chemistry. See Ramsay (1975): Van't Hoff-Le Bel Centennial, p. 74, and Mund (1999): Struktur, Konfiguration, p. 120.
29 Ramsay (1981): Stereochemistry, p. 79.
30 Van't Hoff (1877): Lagerung der Atome im Raume, Braunschweig, p. 3.
31 Ibid., p. 5f.

and 90s.[32] In particular, Emil Fischer was one of the first scientists who started a program on the systematic synthesis of the optical isomers of the sugar *hexose*.[33] One of the goals of this program was to classify sugars and their reaction products according to their stereochemical features, and to develop new experimental methods for the isolation and purification of carbohydrates. In his article on the optical isomers of glucose in 1891,[34] Fischer integrated van't Hoff's theory and presented his findings (illustrated below in Figure 3) as a guideline for the synthesis of different sugar forms. Following van't Hoff's rule, Fischer expected to synthesize sixteen different substances of hexose ($C_6H_{12}O_6$). He assumed that these sixteen substances would be indistinguishable in their chemical structure,[35] but different with respect to the way in which the atoms were spatially arranged in the molecule.[36]

The historian Peter Ramberg points out that Fischer's usage of van't Hoff's theory was purely pragmatic and, moreover, embedded in a complex empirical and classificatory system.[37] He is particularly critical of previous interpretations of Fischer's work on the sugars as an intended proof of van't Hoff's stereochemical theory, i.e. by Claude Hudson and Frieder Lichtenthaler. Against these authors, Ramberg argues that experimental practice and especially the complex methodology involved in the treatment of sugars played a much more important role in Fischer's sugar program than the confirmation or proof of stereochemical theories. He finally claims that Fischer used van't Hoff's theory only for systematic purposes – as "an isomer-counting device" – and not for the actual practice of sugar synthesis.[38]

---

32 Mund (1999): Struktur der Kohlenhydrate, p. 123.
33 Lichtenthaler (1994): Hundert Jahre, p. 2461.
34 Fischer (1891): Ueber die Konfiguration des Traubenzuckers und seiner Isomeren, in: Berichte (24), pp. 1836-1845. Re-published in: Fischer (1906): Untersuchungen über Kohlenhydrate und Fermente, pp. 417-427.
35 Chemical structure refers here to the kind and number of atoms in a molecule, and to their valency (combining capacity). See Rocke (2010): Image and Reality. Kekulé, Kopp, and the scientific imagination, Chicago, p. 13.
36 Fischer (1891): Ueber die Konfiguration, in: Berichte (24), p. 1836.
37 Ramberg (2003): Chemical Structure, p. 246.
38 Ibid.

Figure 3

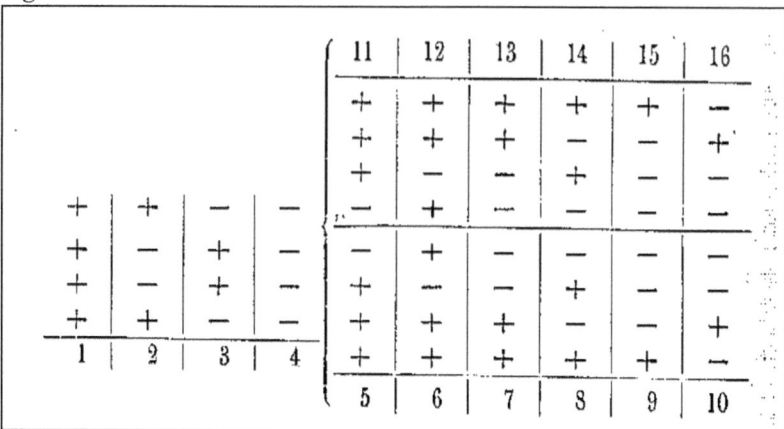

The chart shows the possible molecular arrangements of the sugar hexose to which the structural formula $C_6H_{12}O_6$ is ascribed. The +/- notation represents the relation of the configurations of the isomers. In: Van't Hoff (1877): Lagerung, p. 11, and in: Fischer (1891): Über die Konfiguration des Traubenzuckers und seiner Isomere, p. 1836.

I agree with Ramberg that van't Hoff's theory was at first neither an outstanding, nor an important part of Fischer's investigations in the sugar group. In the beginning of his research on the sugars, from 1886 to 1890,[39] Fischer barely mentioned stereochemical theories, and used the concepts of molecular geometry very sparingly.[40] Rather, his work was marked by a

---

39 Here, I refer to Fischer's work on osazones, especially phenylhydrazine, and their relation to the sugars. See e.g. Fischer (1888): Über die Verbindung des Phenhylhydrazins mit den Zuckerarten IV, in: Berichte (21), pp. 2631-2634, and Id. (1889): Über die Verbindung des Phenhylhydrazins mit den Zuckerarten V, in: Berichte (22), pp. 87-97.

40 An examination of Fischer's published work on the sugars from 1888 to 1894 shows that though he mentioned stereochemical theories and concepts several times in his earlier articles on the sugars, he did not use them for the interpretation of his empirical work or as a means for classification before 1891. (See e.g. Fischer (1888): Über die Verbindung des Phenhylhydrazins mit den Zuckerarten IV, in: Untersuchungen I, p. 162f.; Id. (1890): Über die optischen Isomere des Traubenzuckers, der Gluconsäure und der Zuckersäure, p. 375f.; Id. (1890): Synthesen in der Zuckergruppe I, p. 18). This confirms Ramberg's claim that

2. The lock-and-key analogy in Fischer's research on fermentation | 49

strong empiricism; he neglected speculation as an element of chemical analysis, and concentrated on creating and perfecting laboratory methods. His main goal was the accessibility of the sugars: their isolation, identification, classification and finally their synthesis.[41] In 1891, in his first paper "on the configuration[42] of glucose and its isomers",[43] however, the conviction that sugars differed in their molecular arrangement seems to have laid the groundwork for his motivation to classify them in the first place. Fischer introduced his first article on this topic as follows:

"All previous observations in the sugar group fit so perfectly to the theory of the asymmetric carbon atom that one could already take the venture and try to use the theory as a basis for the classification of these substances."[44]

Yet, what Fischer left unmentioned here is the fact that one year prior to this article he had already classified most of the sugars according to their

---

Fischer's sugar program did not initially aim at a proof or a deduction of the configuration of glucose. (See Ramberg [2003]: Chemical structure, pp. 243-272). However, Fischer's focus shifted in 1891; from there on, he treated configuration as a crucial property of organic molecules, and attempted to classify the previously synthesized sugars according to this criterion. (See Fischer [1891]: Ueber die Konfiguration, in: Untersuchungen über Kohlenhydrate, p. 417). This more or less sudden change of interest remains unexplained in the historical literature.

41 Fischer: Über die Verbindung des Phenylhydrazins mit den Zuckerarten I-IV, in: Untersuchungen über Kohlenhydrate, pp. 138-166.

42 The term "configuration" was introduced by Aemilius Wunderlich in 1886 and referred to the molecular constitution in consideration of the geometric arrangements of the atoms in a molecule. (See Mund [1999]: Struktur der Kohlenhydrate, p. 229).

43 Fischer (1891): Ueber die Konfiguration des Traubenzuckers und seiner Isomere, in: Berichte (24), pp. 1836-1845; re-published in Fischer (1906): Untersuchungen über Kohlenhydrate, p. 417-427.

44 Fischer (1891): Ueber die Konfiguration, in: Berichte (24), p. 1836; in: Untersuchungen über Kohlenhydrate, p. 417.

relations to their reaction products.[45] Information about the structural similarities and differences between the respective substances, derived from experimentation, formed the basis for the classification in 1890 (see figure 4).[46]

Fischer approached the same task again in 1891; this time, however, he included information concerning the configuration of the substances and introduced configuration as one of the basic categorizing criteria.[47] He described the process that led him to the new classification as a systematic derivation from van't Hoff's theory and his chart of the isomeric forms of hexose.

Following Fischer's articles on the sugar group up to this point, it becomes clear that the 1891 article basically served to integrate new findings and relatively sound theoretical assumptions about the role of spatial configuration in the former classification system. It is important to note, though, that the resulting charts and formulas provided only relational information about the configurations of the respective substances, that is, they described the arrangement of e.g. d-glucose in relation to the arrangement of l-glucose (see figure 5).[48] Hence, these representations did not provide an answer to the question of the exact configuration of glucose or to what glucose would look like when considered in a three-dimensional space. Rather they gave rise to the way in which the configuration of glucose was related to the configuration of another sugar or to one of its derivatives.

At the end of 1891, Fischer published a second paper on the configuration of glucose in which he stated that van't Hoff's considerations "formed the basis" of his deductions.[49] As mentioned before, Fischer worked on a

---

45 Fischer (1890): Synthesen in der Zuckergruppe I, in: Untersuchungen über Kohlenhydrate, p. 23.
46 Ibid., p. 18ff.
47 Fischer (1891): Ueber die Konfiguration des Traubenzuckers, in: Berichte (24), p. 1836f; in: Untersuchungen über Kohlenhydrate, p. 417.
48 D- and l-glucose are optical isomers. "D" and "l" stand for the right-handed and the left-handed form of glucose.
49 Fischer (1891): Ueber die Konfiguration II, in: Berichte (24); p. 2684; in: Untersuchungen über Kohlenhydrate, p. 427. ("In den allgemeinen Betrachtungen von van't Hoff, welche meinen speziellen Deduktionen zu Grunde liegen, wird das

2. The lock-and-key analogy in Fischer's research on fermentation | 51

Figure 4

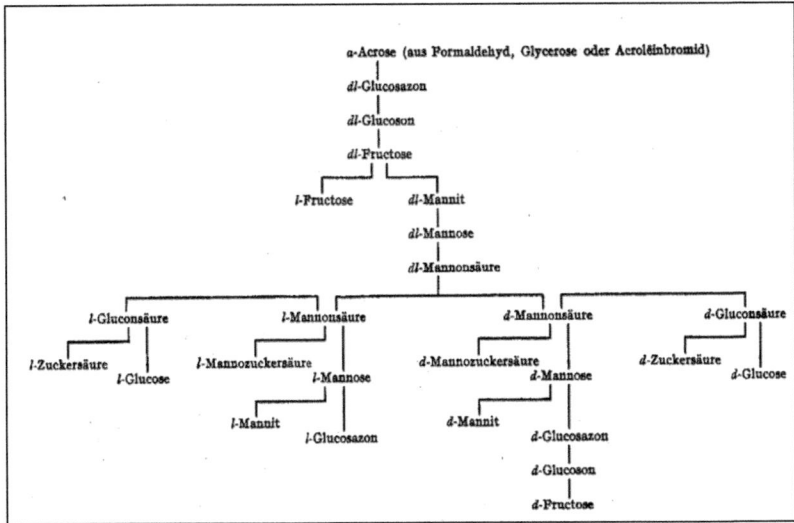

Chart of the synthesized sugars and their derivatives, in: Fischer (1890): Synthesen in der Zuckergruppe I, in: Untersuchungen über Kohlenhydrate, p. 23.

Figure 5

Für die Verbindungen der Hexosegruppe ergeben sich dann, wie leicht ersichtlich, folgende Configurationen:				
Aldosen: COH . CH(OH) . CH(OH) . CH(OH) . CH(OH) . CH$_2$OH				
d. Glucose	−	+	+	+
l. Glucose	+	−	−	−
d. Gulose	+	+	+	−
l. Gulose	−	−	−	+
d. Mannose	+	+	+	+
l. Mannose	−	−	−	−

This table shows one group of the synthesized sugars, namely hexose and its isomers, in: Fischer (1891): Ueber die Konfiguration I, in: Berichte (24), p. 1840.

Kohlenstoffatom 1 nur mit 4 und ebenso das Kohlenstoffatom 2 nur mit 3 verglichen. Im Traubenzucker ist mithin die Anordnung von Wasserstoff und Hydroxyl bei Kohlenstoff 1 umgekehrt wie bei 4, ferner ist sie bei 2 und 3 gleich.")

classification of the sugar group long before he introduced van't Hoff's theory so famously.⁵⁰ This supports Ramberg's claim that Fischer's program on sugars was not initially directed towards an empirical proof of van't Hoff's theory. In fact, Fischer used van't Hoff's theory in a second step to add a new dimension to the study of carbohydrates and to arrive at a more fine-grained picture of the chemical relations between the synthesized sugars and their derivatives. Yet, the conclusion, that this fact counts as compelling evidence for the subsidiary role of van't Hoffs theory in Fischer's sugar program, is premature. Stereochemical theories played a crucial role in Fischer's papers by 1891, even if they did not initially guide him in the attempt to classify and synthesize the sugars. One important contribution that Ramberg mentions is the use of van't Hoff's theory and stereochemical considerations in general as a tool for the systematization of Fischer's empirical material and his methodological approach. But, as Ramberg notes, Fischer only used van't Hoff's theory of the asymmetric carbon atom as a starting point in the elaboration of his own semi-empirical system that was, on the one hand, based on laboratory experience and previously established classifications, but also on the use of theoretical and material molecular models.[51] By 1891, Fischer concentrated his efforts on the development of convenient stereochemical notations which could easily be implemented in the daily practice of organic chemistry. These notations which are still used in contemporary chemical research and education became known as the Fischer-projections (in the following I will refer to these as 'projections').[52] In the next section, I will have a look at the construction process of these stereochemical formulas. It will be shown that by developing the projections, Fischer provided a construction plan, not only for sugar synthesis, but also for chemical synthesis in general.

Another aspect which has so far not been mentioned in the historical literature is the rhetorical and justificatory role of van't Hoff's theory in

---

50 See Fischer (1890): Über die optischen Isomeren des Traubenzuckers, der Gluconsäure und der Zuckersäure, in: Untersuchungen I, p. 362-377, here p. 376; Id. (1890): Synthesen in der Zuckergruppe I, in: Untersuchungen über Kohlenhydrate, pp. 18-20.
51 Ramberg (2003): Chemical Structure, p. 248.
52 Schrader and Rademacher (2009): Kurzes Lehrbuch der Organischen Chemie, Berlin, p. 250.

Fischer's program. In his two articles on the configuration of glucose in 1891, and in "Synthesen in der Zuckergruppe II" in 1894, van't Hoff's theory of the asymmetric carbon atom served as a powerful rhetorical tool in the presentation of chemical synthesis as a systematic enterprise guided by experimental know-how, but also by careful theoretical consideration.[53] As will be shown in section 2.3, this narrative gained further importance in Fischer's program on fermentation and finally played a crucial role in promoting the overall importance and reliability of chemical synthesis.

2.1.2 Studying molecular configuration by means of three-dimensional modeling

So far, I have only loosely characterized the concept of molecular configuration in terms of the three-dimensional arrangement of atoms in a molecule. Fischer's understanding of molecular configuration and its role for organic chemistry was strongly shaped by his approach to sugar synthesis, and articulated most clearly in his two articles on "the configuration of glucose and its isomers" in 1891 and in his second article on the synthesis of sugars in 1894.[54] As mentioned earlier, in all three articles Fischer presented van't Hoff's theory of the asymmetric carbon atom as a guiding tool for his classification of the sugar group. The 1894 article on sugar synthesis marked another turning point in Fischer's approach and can be seen as the first published source in which Fischer introduced his "model-projections", a new kind of structural formulas that were used to represent the configura-

---

[53] Fischer (1891): Ueber die Konfiguration des Traubenzuckers und seiner Isomere, in: Berichte (24), pp. 1836-1845, here p. 1836, p. 1841, and p. 1845. See as well Id. (1891): Ueber die Konfiguration des Traubenzuckers und seiner Isomere II, in Berichte (24), pp. 2683-2687, here p. 2683f., and Id. (1894): Synthesen in der Zuckergruppe II, in Berichte (27), pp. 3189-3232, here p. 3189, p. 3213, and p. 3215.

[54] Fischer (1891): Ueber die Konfiguration des Traubenzuckers, in: Berichte (24), pp. 1836-1840, and 1845; id. (1891): Ueber die Konfiguration des Traubenzuckers II, in: Berichte (24), pp. 2684f.; id. (1894): Synthesen in der Zuckergruppe II, in: Berichte (27), pp. 3189-3232, here p. 3209-3218.

tional differences (which become only visible in a three-dimensional space) between the isomeric sugars on plain paper.[55]

The main issue that was addressed in the second article on the configuration of glucose in 1891 was the ambiguity of earlier attempts to represent molecular configuration and particularly the difficulties of van't Hoff's +/- notation.[56] Fischer stated that "the designation of arrangement in space as + or − [...] can lead easily in case of such complicated molecules to an erroneous interpretation", and that the solution to this problem could be found in a "more detailed interpretation of the formula."[57] In the course of the article it becomes clear that he arrived at this detailed interpretation through the manipulation of three dimensional, material models.[58] Right in the beginning, he suggested to specify van't Hoff's notation "with the following images."[59] What then followed were detailed instructions of how to construct the new stereo-formulas:

"By means of the convenient Friedländer rubber models one constructs at first the molecules of right handed, left handed and inactive tartaric acid und puts these in such a way on the plane of the paper that the four carbon atoms are arranged in a

---

55 Hudson (1941): Emil Fischer's Discovery of the Configuration of Glucose, in: Journal of chemical education, 18 (8), p. 357. See as well Id. (1948): Historical Aspects of Emil Fischer's Fundamental Conventions for Writing Stereo-Formulas in a Plane, in: Advances in Carbohydrate Chemistry, Vol. 3, pp. 1-22, here p. 2.
56 Fischer (1891): Ueber die Konfiguration II, in: Untersuchungen über Kohlenhydrate, p. 428.
57 Fischer (1891): Ueber die Konfiguration II, translated by Claude S. Hudson (1948): Historical Aspects of Emil Fischer's Fundamental Conventions for Writing Stereo-Formulas in a Plane, in: Advances in Carbohydrate Chemistry, Vol. 3, pp. 1-22, here p. 5.
58 Hudson (1948): Historical Aspects of Emil Fischer's Fundamental Conventions, p. 1f.
59 Fischer (1891): Ueber die Konfiguration II, in: Untersuchungen über Kohlenhydrate, p. 428.

straight line and that the respective hydrogens and hydroxyles are arranged above the plain of the paper."[60]

Figure 6

I.	II.
COOH	COOH
H — C — OH	HO — C — H
HO — C — H	H — C — OH
H — C — OH	HO — C — H
*H — C — OH	HO — C — H
COOH	COOH.

Fischer-Projections of d- und l-Zuckersäure, in: Fischer (1891): Ueber die Konfiguration II, p. 429.

As Fischer notes himself, he constructed the formulas on the basis of the so-called Friedländer-models, material rubber models named after the chemist Paul Friedländer.[61] There are no images available of the Friedländer models, but their design and usage is well explained in a report by the chemist Victor Meyer about "the results and aims of stereochemical research" from 1890.[62] According to Meyer, the material models consisted of

---

60 Ibid. ("Mit Hilfe des Modells erkennt man leicht, dass beim Kohlenstoffatom 2 das Zeichen wechselt, je nachdem, ob man es mit 1 oder 3 vergleicht. Da also der Ausdruck [anhand der +/- Bezeichnung; R.M.] für die Konfiguration des Traubenzuckers zweideutig ist, so scheint es mir zweckmäßig, denselben durch folgende Bilder zu verdeutlichen. Man konstruiere zunächst mit Hilfe der so bequemen Friedländer'schen Gummimodelle die Moleküle der Rechtsweinsäure, Linksweinsäure und inaktiven Weinsäure und lege dieselben derart auf die Ebene des Papiers, daß die vier Kohlenstoffatome in einer geraden Linie sich befinden und daß die in Betracht kommenden Wasserstoffe und Hydroxyle über die Ebene des Papiers stehen.")
61 Fischer (1891): Ueber die Konfiguration II, p. 428. See as well Meyer (1890): Ergebnisse und Ziele der stereochemischen Forschung, in: Berichte der deutschen chemischen Gesellschaft (23), p. 571f.
62 Meyer (1890): Ergebnisse, p. 571.

thin rubber sleeves that were brazed in the middle; univalent atoms were represented by means of small wooden sticks whose cone ends had different colors.[63] They were available as combinatory tool-kits, allowing the user to frequently adjust his molecular models according to new experimental observations or theoretical refinements.[64]

In the UK and Germany, three-dimensional wooden, rubber and card board models were an integral part of chemical education in the second half of the 19th century. According to Christoph Meinel chemists prepared those models "for teaching purposes, and in using them they learned to link the mind's eye with theoretical notions, with the manipulating hand, and with laboratory practice."[65]

It is thus likely that Fischer's instructions of how to use the Friedländer models referred to a common practice that chemists who were trained in the mid-and late 19th century could relate to. Yet despite the common use of material modeling for educational purposes, modeling practice was not something chemists frequently wrote about in journal articles.[66] Van't Hoff was one of the few and early exceptions who emphasized that, in addition to their pedagogical advantages, material models could be used for epistemic problem-solving.[67]

Fischer's contributions to the existing practice of material modeling can be summarized as follows: First, he introduced new means of representation and model manipulation that allowed for a three-dimensional perspective on molecules on plain paper. Key to this practice was the adjustment of three-dimensional objects to new findings and hypotheses as well as the

---

63 Ibid., p. 572. ("Sie bestehen aus Gummischläuchen, deren lichte Weite etwa der Dicke eines Zündholzes gleich ist, und die in der Mitte zusammengelöthet sind; sie erstrecken sich in derselben Weise, wie die Drähte an dem älteren [Kugel-Stab; R.M.] Modell, in den Raum. Einwerthige Atome werden angedeutet durch kleine Hölzchen mit verschiedenfarbigen Spitzen, welche die Form von Zündhölzern haben. Sollen Derivate der C2-Reihe dargestellt werden, so steckt man ein Hölzchen in den Schlauch hinein, lässt es zur Hälfte herausragen und befästigt daran ein zweites Atommodell.").
64 Ibid.
65 Meinel (2004): Molecules and Croquet Balls, here p. 265.
66 Meinel (2009): Kugel und Stäbchen, p. 257.
67 Ibid., p. 265.

implementation of this knowledge in a chemical notation and classification system. The formulas themselves could not depict the arrangements of atoms in three-dimensional space, but the material rubber models which they referred to could.[68] A three-dimensional perspective was provided by additional instructions of how to construct, manipulate and interpret these novel "paper tools."[69] Secondly, Fischer called attention to a common, but hidden practice of model construction by spelling out how to use and manipulate material models. Given Fischer's high reputation as an established scientist and teacher, it is likely that his way of introducing the projections conveyed the message that working with models and actually using them for epistemic purposes was officially accepted. Thirdly, apart from their contribution to new ways of representation in late 19th century chemistry, the projections strengthened the position that molecular configuration was not only an important theoretical, but as well an empirical factor in the study of organic substances.

2.2 ENVISIONING NEW POSSIBILITIES OF CHEMICAL SYNTHESIS FOR BIOLOGY AND MEDICINE

Envisioning future possibilities of biological research and medical practice by means of chemical synthesis became a key element in Fischer's reports. A lot of his articles started or ended with a forecast, giving rise to possible

---

68 For a more detailed depiction of the usage of material models in 19th century organic chemistry see Meinel (2004): Molecules and croquet balls, in: Chadarevian and Hopwood (eds.): Models: The Third Dimension of Science, pp. 242-276. See as well Francoeur (2000): Beyond dematerialization and inscription, in: HYLE – International Journal for Philosophy of Chemistry, Vol. 6, No. 1, pp. 63-84, here p. 63.

69 Ursula Klein has coined the term "paper tools" for the constructive and autonomous role of chemical formulas in the 19th century. (See Klein (2003): Experiments, models, paper tools, Stanford, p. 35; Ead. (2001): The Creative Power of Paper Tools in Early Nineteenth Century, in: Ead. (ed.): Tools and Modes of Representation in the Laboratory Sciences, Dordrecht, p. 13ff.).

applications of his synthetic achievements.[70] He often pointed to the importance of his achievements in carbohydrate chemistry for biology, medicine, and the chemical industry. Between 1890 and 1898, the number of comments he made on the usefulness of not only the new, synthesized products but also of the synthetic methods and stereochemical theories increased.[71] One of the most famous of these visionary passages is the following, in which he leads over to the question of sugar production occurring in natural organisms, i.e. plants and animals. The pressing issue that Fischer addresses here is the intervention in natural sugar production by means of chemical synthesis. He enthusiastically states:

"If it was now possible, to feed the assimilated plant […] with a different composed sugar, the plant might be forced to even produce a different protein. And should we not expect that this modified building material [Baumaterial] would cause a modification of the architecture? We would then be able to chemically intervene on the constitution of the organism, and this might lead to the strangest appearances, to modifications of the form which leave everything behind, hitherto achieved by means of breeding and crossing. Biology is confronted with a question that, as far as I know, has not been addressed yet, and which could have not been addressed before chemistry provided the material for this experiment."[72]

---

70 Mazumdar (1998): Species, p. 193.
71 Ibid.
72 Fischer (1890): Synthesen in der Zuckergruppe I, in: Berichte (23), pp. 2114-2141, here p. 29. ("Wenn es nun möglich wäre, die assimilierende Pflanze […] durch einen anders zusammengesetzten Zucker zu ernähren, so würde sie vielleicht gezwungen, sogar ein anderes Eiweiß zu bilden. Und dürfen wir dann nicht erwarten, daß das veränderte Baumaterial eine Veränderung der Architektur zur Folge hat? Wir würden so einen chemischen Einfluß auf die Gestaltung des Organismus gewinnen und daß müsste zu den sonderbarsten Erscheinungen führen, zu Veränderungen der Form, welche alles weit hinter sich lassen, was man bisher durch Züchtung und Kreuzung erreicht hat. Die Biologie steht hier vor einer Frage, welche meines Wissens bisher nicht aufgeworfen wurde, welche auch in dieser Form nicht aufgeworfen werden konnte, bevor die Chemie das Material für den Versuch geliefert hat.").

2. The lock-and-key analogy in Fischer's research on fermentation | 59

Pauline Mazumdar points out that this "kind of chemical-biological idea was quite Fischer-specific."[73] Fischer garnished a lot of his experimental achievements by envisioning the future and utility of synthetic chemistry, mostly for biological, medical and industrial purposes. In "The chemistry of carbohydrates and its role for physiology" ("Die Chemie der Kohlenhydrate und ihre Bedeutung für die Physiologie"),[74] he emphasized once more the importance of synthetic chemistry for biology and medicine:

"We have come to an end in the synthetic field, and some of you might have wondered, why I invited a group of physicians to examine this topic in such detail. Indeed, I would have thought of this venture as daring, if this would only concern objects of chemical interest, if there was no hope that the results of this synthetic work would soon influence biological research [...]. To what extent this seems to be possible shall be demonstrated by the following considerations."[75]

This quote allows an interesting perspective on the role Fischer attributed to synthetic chemistry. It raises two dimensions in which chemical synthesis is important for biology and for economic welfare. First, it provides the material for further investigations and for industrial growth. Secondly, Fischer emphasizes that his achievements in the sugar group raise the question of intervention on the form of an organism for the first time. He even goes one step further by claiming that biology would not have been able to approach

---

73 Mazumdar (1995): Species, p. 193.
74 Fischer (1909 {1894}): Die Chemie der Kohlenhydrate und ihre Bedeutung für die Physiologie, in: Untersuchungen über Kohlenhydrate und Fermente, pp. 96-115.
75 Fischer (1909 {1894}): Die Chemie der Kohlenhydrate und ihre Bedeutung für die Physiologie, in: Untersuchungen über Kohlenhydrate und Fermente, p. 107. ("Wir sind am Ende des chemischen Gebietes angelangt, und mancher von Ihnen mag sich erstaunt gefragt haben, was mich veranlassen konnte, eine Versammlung von Ärzten zu so eingehender Betrachtung desselben einzuladen. In der Tat würde mir selbst solch Unternehmen gewagt erschienen sein, wenn es sich hier um ausschließlich chemische Dinge handelte, wenn nicht die Hoffnung bestände, dass die Ergebnisse der synthetischen Arbeit über den Rahmen der Fachwissenschaft hinaus bald die biologische Forschung beeinflussen würden. Wie weit das möglich erscheint, sollen die folgenden Betrachtungen zeigen.").

this issue without the help of synthetic chemistry. It is important to note, however, that synthesis could only fulfill this role because it was more than just the production of substances. Rather, it provided a systematic methodology for future projects.

These visionary passages are especially interesting with respect to Fischer's image, as he was known for his careful examination of hypotheses, as well as for his skepticism towards theory driven research. His students and colleagues portrayed him as an empiricist,[76] despite his obvious support of stereochemical theories in the absence of clear empirical evidence. Emil Abderhalden, one of Fischer's students who himself would later became quite influential in protein chemistry and physiology,[77] wrote in his commemoration of Fischer that

"[h]e did not mystify his work with daring speculations. He was a man of actuality [Wirklichkeitsmensch]. For him only facts counted. I still remember his chemistry lectures in which he presented the heroic efforts of carbohydrate chemistry and explained to the breathlessly listening audience how the Le Bel/van't Hoff theory triumphed as applied to the carbohydrates and especially to the hexoses, in fact, what he did not mention, only due to his masterly syntheses. All eyes repeatedly turned to the configuration formulas of the hexoses."[78]

---

76 See Hoesch (1921): Emil Fischer. Sein Leben und sein Werk, in: Berichte der deutschen chemischen Gesellschaft, Sonderheft des 54. Jahrgangs, p. 95.
77 See Fruton (1985): Contrasts in Scientific Style, p. 338.
78 Abderhalden (1919): Die Bedeutung von Fischers Lebenswerk für die Physiologie, in: Die Naturwissenschaften, 7 (46), pp. 860-868, here p. 860. ("Er umwob seine Arbeit nicht mit kühnen Spekulationen. Er war Wirklichkeitsmensch. Ihm galten nur Tatsachen etwas. Noch erinnere ich mich seiner Vorlesungen über Chemie, in der er der atemlos lauschenden Zuhörerschaft die gewaltigen Errungenschaften der Kohlehydratchemie vorführte und erläuterte, wie die Theorie von Le Bel und van't Hoff bei den Kohlehydraten und insbesondere Hexosen Triumpfe gefeiert hatte, und zwar, was er nicht zum Ausdruck brachte, einzig durch seine meisterhaften Synthese. Aller Augen wandten sich immer wieder auf die Konfigurationsformeln der Hexosen hin.").

2. The lock-and-key analogy in Fischer's research on fermentation | 61

This passage from Abderhalden's lecture at Fischer's wake[79] exemplifies that Fischer's work on the sugars lasted in the heads of his successors. It also points to the tension in the perception of Fischer as "a man of actuality" who avoids speculations on the one hand and, on the other hand, as an advocate of stereochemical theories which had strong speculative elements.

To summarize, Fischer contributed to implementing the conception of molecular geometry in chemical practice by demonstrating that this conception can serve as a powerful tool for empirical investigations, e.g. for sugar synthesis and classification, and also as a guiding tool for experiments in fermentation chemistry. It helped to envision new possibilities of biological and medical research driven by the methods of chemical synthesis and stereochemical concepts. If biological functions of organisms could somehow be determined by molecular geometry, then theories and methods of organic chemistry would play a crucial role in solving biological and medical problems systematically.

---

79 Ibid., p. 863. ("Emil Fischer hat auf die Physiologie und alle mit ihr zusammenhängenden Gebiete vor allem in zwei Richtungen tief gehenden Einfluss ausgeübt. Auf der einen Seite hat er durch Schaffung eines großen, tief durchdachten und mit größtem experimentellen Geschick errungenen Tatsachenmaterials für eine große Zahl von Fragestellungen die Fundamente geliefert. [...] Auf der anderen Seite verdanken wir Emil Fischer - und daß ist leider viel zu wenig bekannt – Vorstellungskomplexe, die für gewaltige Forschungsgebiete richtungsgebend geworden sind. Der klare Begriff der Spezifität, der spezifischen Wirkung, ist erst durch seine fundamentellen Arbeiten über die Beziehungen zwischen der Struktur und der Konfiguration eines Substrates und bestimmten Fermentwirkungen möglich geworden. Alle späteren Anschauungen über spezifische Einstellungen von Antitoxin auf Toxin, kurz die Seitenkettentheorie nahm ohne jeden Zweifel ihren Ausgangspunkt von den genannten, so fruchtbaren Vorstellungen Emil Fischers. Es ist von höchstem Interesse, daß die letzte Arbeit, die eben jetzt als neuer Zeuge von Emil Fischers noch ganz ungebrochener Arbeitskraft und höchster geistiger Regsamkeit erschienen ist, sich wieder mit den grundlegenden Fragen der spezifischen Wirkung, und zwar speziell am Beispiel des Emulsins beschäftigt.").

2.3 DISCOVERING THE STEREOCHEMICAL MECHANISM OF FERMENTATION

After his success in carbohydrate chemistry, Fischer was confident that the three-dimensionality of molecules was essential, not only for the physical and chemical behavior of chemical substances, but also and most importantly for their biological functions in the organism.[80] This confidence was strengthened by the possibility of creating formulas that made the configurational differences of the sugars visible and that were sometimes treated as a systematic proof of the concept of configuration.[81] He was certain by now that the existence of isomeric substances was related to differences in the spatial arrangement of atoms in a molecule. Based on his experiences, he formulated the hypothesis "that a sugar would only be attacked and fermented, if the geometrical structure of its molecule is not too far from that of the protein" and compared this kind of reaction with the relation between a lock and a key.[82]

Between 1890 and 1894, Fischer had already emphasized several times that studies on the geometrical arrangement of carbohydrates were of the "greatest importance" for biology and that a stereochemical perspective would open a whole new set of questions for the investigation of biological phenomena.[83] Like many other organic chemists who became interested in biological phenomena, Fischer focused on the respective substances and their characteristics, rather than on their effects in the organism or on the organism itself.[84] Hence, although his work on the structure of biologically

---

80 Mazumdar (1995): Species, p. 193.
81 See e.g. Hudson (1941): Emil Fischer's Discovery of the Configuration of Glucose; Id. (1948): Historical Aspects of Emil Fischer's Fundamental Conventions, and Lichtenthaler (1994): Hundert Jahre.
82 Fischer (1894): Synthesen II, in: Untersuchungen über Kohlenhydrate, p. 108f.; translated by Mazumdar (1995): Species, p. 198.
83 Fischer (1894): Synthesen II, in: Untersuchungen über Kohlenhydrate, p. 108; translated by Mazumdar (1995): Species, p. 194.
84 Nonetheless, as Mazumdar mentions, Fischer "stood out among the organic chemists in that his interests centered on the structure of the substances found in organisms, that is on structural biochemistry rather than pure organic chemistry." (Mazumdar [1995]: Species, p. 191). For a more detailed analysis of Fisch-

important substances (such as carbohydrates and proteins) was often perceived as one of the foundations for the establishment of the biochemical sciences, Fischer's approach was at its core a chemical one.[85] Despite its clear chemical focus, however, the significance of Fischer's fermentation program was widely regarded among biologists, and it became known as one of the first programs that explicitly linked molecular geometry and biological function.[86]

Together with his colleague Hans Thierfelder, Fischer published an article on the chemical reactions between different sorts of yeast and a number of stereoisomeric sugars that he had hitherto classified and synthesized.[87] Their experimental observations led to the assumption that the cells of yeast could only react with those sugars that showed a related chemical configuration.[88] In the same year Fischer conducted another series of exper-

---

er's laboratory style, see as well Fruton (1985): Contrasts in Scientific Style. Emil Fischer and Franz Hofmeister: Their Research Groups and Their Theory of Protein Structure, in: Proceedings of the American Philosophical Society, Vol. 129, No. 4, p. 317f. and 326f.

85 Fruton (1985): Contrasts, p. 317.
86 Gilbert and Greenberg (1984): Intellectual Traditions, p. 26, p. 30, and p. 32; Barnett and Lichtenthaler (2002): A history of research on yeast 3: Emil Fischer, Eduard Buchner and their contemporaries, 1880-1900, p. 372; Morange (1998): A history of molecular biology, p. 12.
87 Fischer und Thierfelder (1894): Verhalten der verschiedenen Zucker gegen reine Hefen, in: Berichte (27), pp. 2031-2037.
88 Fischer (1894): Einfluss der Configuration auf die Wirkung der Enzyme, in: Berichte (27), p. 2985f. ("Das verschiedene Verhalten der stereoisomeren Hexosen gegen Hefe hat T h i e r f e l d e r und mich zu der Hypothese geführt, dass die activen chemischen Agentien der Hefezelle nur in diejenigen Zucker eingreifen können, mit denen sie eine verwandte Configuration besitzen. Diese stereochemische Auffassung des Gährprocesses musste an Wahrscheinlichkeit gewinnen, wenn es möglich war, ähnliche Verschiedenheiten auch bei den vom Organismus abtrennbaren Fermenten, den sogenannten Enzymen, festzustellen. Das ist mir nun in unzweideutiger Weise zunächst für zwei glucosidspaltende Enzyme, das Invertin und Emulsin, gelungen. Das Mittel dazu boten die künstlichen Glucoside, welche nach den von mir aufgefundenen Verfahren aus den verschiedenen Zuckern und den Alkoholen in grosser Zahl bereitet werden können. Zum

iments in which he concentrated on enzymes which could be isolated from the yeast organism.[89] These advanced studies served to support the hypothesis that the tested reactants of yeast and sugar have related configurations and thereby strengthened the stereochemical perspective on the fermentation process. Fischer gained promising results with two enzymes, *Emulsin* and *Invertin*. He observed that the various sorts of sugars were fermented by the respective enzyme in different ways, e.g. regarding the time that was needed for the enzyme to dissolve the sugars.[90] In some cases, e.g. in experiments with glycosides, Fischer observed that there was no such reaction between enzymes and sugars – both emulsin and invertin did not have any effect on the respective sugars.[91] The distinguishing feature of the appearance and intensity of the reaction between enzymes and sugars, following Fischer, was the configuration of the sugars, i.e. their geometrical molecular arrangement. He claimed that experiments with emulsin and invertin gave reason to generally assume that enzymes are highly selective with respect to the configurations of their dissolving products.[92] "The limited ef-

---

Vergleich wurden aber auch mehrere natürliche Producte der aromatischen Reihe und ebenso einige Polysaccharide, welche ich als die Glucoside der Zucker selbst betrachte, in den Kreis der Untersuchung gezogen. Das Ergebniss derselben lässt sich in den Satz zusammenfassen, dass die Wirkung der beiden Enzyme in auffallender Weise von der Configuration des Glucosidmoleküls abhängig ist.").

89 Ibid., p. 2986.
90 Ibid.
91 Ibid., p. 2988.
92 Ibid., p. 2993. ("Aber schon genügen die Beobachtungen, um prinzipiell zu beweisen, das [sic.] die Enzyme bezüglich der Konfiguration ebenso wählerisch sind wie die Hefe und andere Mikroorganismen. Die Analogie beider Phänomene erscheint in diesem Punkte so vollkommen, daß man für sie die gleiche Ursache annehmen darf und damit kehre ich zu der vorher erwähnten Hypothese von Thierfelder und mir zurück. Invertin und Emulsin haben bekanntlich manche Ähnlichkeit mit den Proteinstoffen und besitzen wie jene unzweifelhaft ein asymmetrisch gebautes Molekül. Ihre beschränkte Wirkung auf die Glucoside ließe sich also auch durch die Annahme erklären, daß nur bei ähnlichem geometrischen Bau diejenige Annäherung der Moleküle stattfinden kann, welche zur Auslösung des chemischen Vorgangs erforderlich ist. Um ein Bild zu gebrau-

fect of enzymes on glycosides could thus be explained by the hypothesis that the respective molecules are only capable of approximation" when their configurations are similar.[93] He further noted that the approximation of enzymatic and sugar molecules was a necessary condition for their chemical reaction to happen. The passage that followed has since been emphasized many times by scientists and historians and has often been designated as a breakthrough in the early history of biochemistry.[94] "To make use of an image", Fischer wrote, "I shall say that enzyme and glycoside must fit each other like lock and key in order to have any chemical effect on each other."[95] By using this analogy, Fischer emphasized that the respective units have a relationship which is similar to the one between a lock and a key, but left open how exactly enzyme and glycoside have to fit to cause a chemical reaction; Fischer did not know what this would look like in detail. At this time, there was little known about the structure of enzymes and no one was capable of synthesizing them.[96] One of the few alternatives left to

---

chen, will ich sagen, daß Enzym und Glucosid wie Schloß und Schlüssel zueinander passen müssen, um eine chemische Wirkung aufeinander ausüben zu können. Diese Vorstellung hat jedenfalls an Wahrscheinlichkeit und an Wert für die stereochemische Forschung gewonnen, nachdem die Erscheinung selbst aus dem biologischen auf das rein chemische Gebiet verlegt ist. Sie bildet eine Erweiterung der Theorie der Asymmetrie, ohne aber eine direkte Konsequenz derselben zu sein; denn die Überzeugung, daß der geometrische Bau des Moleküls selbst bei Spiegelbildformen einen so großen Einfluß auf das Spiel der chemischen Affinitäten ausübe, konnte meiner Ansicht nach nur durch neue tatsächliche Beobachtungen gewonnen werden.").

93 Fischer (1894): Einfluss der Configuration, p. 2993.
94 See e.g. Gilbert and Greenberg (1984): Intellectual Traditions, p. 18.; Lichtenthaler (1994): Hundert Jahre, p. 2371; Barnett and Lichtenthaler (2001): A history of research on yeast 3: Emil Fischer, Eduard Buchner and their contemporaries, 1880-1900, in: Yeast, 18, pp. 363-388; Morange (1998): A History, p. 15; and Travis (2008): Models for biological research, pp. 88ff.
95 Fischer (1894): Einfluss der Configuration, p. 2992; translated by Mazumdar (1995), p. 198.
96 Glaesmer (2004): Zur Entwicklung der wissenschaftlichen Verflechtung der Chemie mit anderen Wissenschaften bei der Erforschung von Struktur, Funktion und Synthese von Proteinen im 20. Jahrhundert, Dissertation, Berlin, p. 24.

study the chemical features of enzymes was to study their reactions with substances that were better known and accessible to chemical intervention. For Fischer, the sugars turned out to be the ideal candidates to carry out this project, since his previous studies provided him with information about the structural and stereochemical features of these substances.[97]

Though being aware of the speculative character of the lock-and-key analogy, Fischer pointed out that this speculation gave the impulse to study the influence of molecular configuration on the fermentation of glycosides. In a later article, he emphasized that the lock-and-key analogy provided "experimental research with a very specific and attackable problem", namely to search for *stereochemical differences* in the fermentation of glycosides.[98] The experiments with emulsin and invertin also led Fischer to the

---

[97] Fischer (1894): Einfluss der Configuration, p. 2986.

[98] Fischer (1909 {1898}): Bedeutung der Stereochemie für die Physiologie, in: Untersuchungen über Kohlenhydrate und Fermente (1884-1908), p. 134. ("Der Grund dieser Erscheinungen liegt aller Wahrscheinlichkeit nach in dem asymmetrischen Bau des Enzymmoleküls. Denn wenn man diese Stoffe auch noch nicht in reinem Zustand kennt, so ist ihre Ähnlichkeit mit den Proteinstoffen doch so groß und ihre Entstehung aus letzteren so wahrscheinlich, daß sie zweifellos selbst als optisch aktive und mithin asymmetrisch molekulare Gebilde zu betrachten sind. Das hat zu der Hypothese geführt, daß zwischen den Enzymen und ihrem Angriffsobject eine Ähnlichkeit der molekularen Konfiguration bestehen muss, wenn Reaktion erfolgen soll. Um diesen Gedanken anschaulicher zu machen, habe ich das Bild von Schloß und Schlüssel gebraucht. Ich bin weit entfernt, diese Hypothese den ausgebildeten Theorien unserer Wissenschaft an die Seite stellen zu wollen, und ich gebe gern zu, daß sie erst dann eingehend geprüft werden kann, wenn wir imstande sind, die Enzyme im reinen Zustand zu isolieren und ihre Konfiguration zu erforschen. Trotzdem halte ich gegenüber den Einwänden von Bourquelot und Duclaux eine solche Spekulation nicht für unstatthaft, wenn sie auch den Tatsachen vorauseilt. Denn sie hat mich veranlasst, die bei der alkoholischen Gärung der Monosaccharide gemachten Erfahrungen bei den Glucosiden zu verfolgen; sie stellt der experimentellen Forschung weiter das ganz bestimmte und angreifbare Problem, dieselben Unterschiede, welche wir in der enzymatischen Wirkung beobachten, bei einfacheren, asymmetrisch gebauten Substanzen von bekann-

conclusion that the dependency of the enzyme-substrate reaction on molecular geometry is not the same as the one Pasteur pointed to in his research on optical antipodes. In his Faraday lecture in 1907, Fischer clarified retrospectively that even miniscule differences in the configuration of the organic substrate can influence the reaction between enzyme and substrate.[99] As previously mentioned, Pasteur already speculated in 1864 that the observed differences of structurally equivalent substances with respect to their optical activity could be explained by the geometrical structure of molecules. In his fermentation experiments with mold, Pasteur observed that only one of two substances with the same chemical compounds was optically active. Based on these observations, he suggested that there were two different forms of the same substance and that the differences in their chemical behavior were caused by the match or mismatch of molecules in a substance.[100] The novel aspect of Fischer's stereochemical perspective on fermentation, as opposed to Pasteur's view, lay in the assumption that even the smallest details in the molecular configuration of organic substances could cause a difference in the process of their fermentation.[101] As previously explained, Fischer found a way to make these details visible by means of the model-projections.

---

ter Konstitution aufzusuchen, und ich zweifle nicht daran, daß schon die nächste Zukunft uns hier wertvolle Resultate bringen wird.").

99   Fischer (1924 {1912}): Organische Synthese und Biologie (Faraday Lecture), in: Bergmann (ed.): Untersuchungen aus verschiedenen Gebieten. Vorträge und Abhandlungen allgemeinen Inhalts von Emil Fischer, p. 784. ("Die erweiterte Kenntnis der Monosaccharide ist in mehrfacher Beziehung der biologischen Forschung zugute gekommen. Insbesondere hat sie zu einer Vertiefung unserer Kenntnisse über die Wirkung der Enzyme geführt. Aus dem verschiedenen Verhalten der zahlreichen synthetischen Glucoside gegen Emulsin und die Fermente der Hefe konnte ich den Schluß ziehen, daß hier nicht allein der Gegensatz zwischen zwei optischen Antipoden existiert, den Pasteur in dem Verhalten gegen Schimmelpilze festgestellt hat, sondern daß ganz kleine Differenzen im sterischen Aufbau genügen, um die Wirkung des Enzyms zu verhindern.").

100 Pasteur (1860): Ueber die Asymmetrie, p. 13.

101 Fischer (1924 {1912}): Organische Synthese und Biologie (Faraday Lecture), p. 784.

Yet, as Frieder Lichtenthaler mentions in his anniversary article on the occasion of the lock-and-key analogy's 100th birthday, Fischer did not use the analogy for long.[102] Despite the self-ascribed fruitful effect it had on his research on fermentation,[103] the analogy did not seem to play a role in his second long-term program on the nature and chemical production of proteins. In these studies, Fischer directed all his efforts towards the development of new methods for the synthesis of proteins.[104] He approached this task in the same manner as the synthesis of sugars, that is, by analyzing the derivatives of proteins, the peptones and the amino acids first.[105] Such an analysis was possible due to the process of hydrolysis which could be initiated by acids, alkali or digestive juices (e.g. gastric acids). In living organisms, proteins could be transformed into peptones in the first stage of hydrolysis; in a second stage taking place in the gut of the respective organism, these peptones were once more dissected into amino acids.[106] After Fischer was able to follow the path of protein hydrolysis and synthesize the previously identified amino acids, he stated that

"the artificial construction of proteins seems to amount to the task of linking these amino acids, according to their correct selection and sequence, by the separation of water molecules. In the last five years I have thus tried to find suitable methods to this end and indeed, I succeeded in gaining products [...] which are in many ways similar to proteins."[107]

---

102 Lichtenthaler (1994): 100 Jahre Schlüssel-Schloss, p. 2371.
103 Fischer (1924 {1898}): Bedeutung der Stereochemie für die Physiologie, p. 134.
104 Fischer (1924 {1907}): Proteine und Polypeptide, in: Untersuchungen auf verschiedenen Gebieten, pp. 748-757.
105 Ibid., p. 749.
106 Fischer (1924 {1907}): Proteine und Polypeptide, p. 749.
107 Ibid., p. 751. ("Der künstliche Aufbau der Proteine selbst scheint also im Wesentlichen auf die Aufgabe hinauszulaufen, diese Aminosäuren in richtiger Auswahl und Reihenfolge durch Abspaltung von Wasser miteinander zu verknüpfen. Ich habe mich deshalb seit fünf Jahren bemüht, geeignete Methoden für diesen Zweck aufzufinden und es ist mir in der Tat gelungen, durch die Verkupplung der verschiedenen Aminosäuren Produkte zu gewinnen, die zu-

2. The lock-and-key analogy in Fischer's research on fermentation | 69

One can only speculate about Fischer's motivations and the reasons for the absence of the lock-and-key analogy in this context. However, Fischer's previous emphasis on the fruitfulness of the lock-and-key analogy,[108] and the problems he faced in the course of his protein program suggest that his research on proteins was not sufficiently developed for the usage of such an analogy: It would have suggested too much at this point. As mentioned above, Fischer had a strong distaste for groundless speculation and reminded his students persistently that experimentation should always precede speculative hypothesizing.[109] It is likely that this distaste also kept him from expanding his use of the lock-and-key analogy for the protein program. Fischer's statement from 1898 that "he is not willing to compare" the lock-and-key analogy with more developed chemical theories until the controlled synthesis of substances to which the analogy applies is possible, illustrates this very nicely.[110] My claim that Fischer hesitated to use the analogy in an early state of the protein program can further be supported by the fact that in his work on the sugars he also refrained from theorizing about the molecular arrangement of the respective substances before their synthesis was accomplished.

---

    erst den Peptonen und bei fortgesetzter Synthese den Proteinen sehr ähnlich sind.").
108 Fischer (1924 {1898}): Bedeutung der Stereochemie für die Physiologie, in: Untersuchungen auf verschiedenen Gebieten, p. 134.
109 Hoesch (1921): Emil Fischer, p. 95; Fruton (2002): Methods and Styles in the development of chemistry, in: Memoirs of the American Philosophical Society held at Philadelphia for promoting useful knowledge, Vol. 245, p. 152.
110 Fischer (1924 {1898}): Bedeutung der Stereochemie für die Physiologie, in: Untersuchungen auf verschiedenen Gebieten, p. 134. See also Lichtenthaler (1994): Hundert Jahre, p. 2464.

2.4 THE HEURISTIC ROLE OF THE LOCK-AND-KEY ANALOGY IN FISCHER'S PROGRAM

In what follows, I will argue that Fischer used the lock-and-key analogy in a heuristic way and more specifically that he used it to solve an explanatory research problem by applying certain heuristic strategies. I will at first develop a notion of heuristics and heuristic tools based on philosophical literature on heuristic problem solving. These considerations will then lay the groundwork for the interpretation of the role of the lock-and-key analogy in Fischer's fermentation research. It will also provide an orientation for the analysis of the lock-and-key analogy in 20th century immunology (chapter 3) and molecular biology (chapter 4).

2.4.1 Heuristic strategies in the philosophical literature

The understanding of scientific practice as one that involves the usage of heuristic strategies was supported by philosophers working on the nature of scientific problem-solving in the 1980s and 90s.[111] In particular, Robert

---

[111] See e.g. Bechtel and Richardson (2010 {1993}): Discovering Complexity. Decomposition and Localization as Strategies of Scientific Research, Cambridge (Ma); Richardson (1999): Heuristics and satisficing, in: Bechtel/Graham: A companion to cognitive science, pp. 566-575. Bechtel's and Richardson's claims about the nature of scientific discovery were in part based on Allen Newell's and Herbert A. Simon's conceptions of human problem solving and rationality, but also differed from their approaches with respect to the "bounded rationality" thesis. (See Bechtel and Richardson (2010 {1993}: Discovering Complexity, p. 12). In his 1999 article, Richardson makes clear how this thesis differs from Newell's and Simon's classical conceptions of human problem solving: "If we think of problem solving as a search through the space of possibilities as it was conceptualized by Allen Newell and Herbert A. Simon, limitations on search entail that for all but the simplest problems we can investigate at most a small proportion of the possibilities. Moreover, our evaluation of these possibilities will be uncertain and incomplete. We must rely on heuristic methods for pruning the tree of possibilities and for evaluating the possi-

Richardson and William Bechtel (1993[2010]) have offered one influential philosophical and historical analysis of heuristic problem-solving strategies in scientific discovery processes. In "Discovering Complexity" Bechtel and Richardson focus on the incomplete nature of knowledge generation processes and the importance of heuristic failure for scientific development.[112] The authors exemplify how certain strategies that led to a reduction of the amount of complexity of scientific phenomena fruitfully changed the course of historical experimental research programs. This fruitfulness lay in the insight that episodes of scientific discovery, which implemented heuristic strategizing, were made feasible in the light of uncertainty. They concentrate on two such strategies, "decomposition and localization", which functioned as heuristic means of scientific discovery in several research programs, ranging from 18th century chemistry to late-20th century biology.[113] They call these strategies "heuristics" in order to emphasize that

"they are fallible research strategies. Most simply, we may have misidentified some of the component operations in a mechanism. In other cases that we regard as both common and significant, the very error is in assuming the system is decomposable or even nearly decomposable – i.e., that the phenomenon of interest is due to component operations discretely localized in component parts of a mechanism. In such cases the mechanistic model will not adequately capture the system's behavior, thereby revealing that there is a more complex organization than was initially assumed, one in which there is a high degree of interactivity. [...] In either case, the elaboration of the mechanist model can be a means of discovery."[114]

Bechtel and Richardson go on to draw a picture of science in which doing research consists of making claims about the causes of phenomena that usually turn out to be incomplete or false. The authors defend the claim that knowledge generation is incomplete, and typically involves trial and error procedures. However, they also make it clear that scientists learn to use errors productively in problem solving processes, as long as they develop

---

bilities we consider." (Richardson (1999): Heuristics and satisficing, in: Bechtel/Graham: A companion to cognitive science, Blackwell, p. 566).
112 Bechtel and Richardson (2010 {1993}): Discovering Complexity, p. xvii.
113 Ibid., p. xxviii-xxxvii, and pp. 23-27.
114 Ibid., p. xxx.

heuristic strategies to systematically control them.[115] In his 2007 collection "Re-Engineering Philosophy for Limited Beings" Wimsatt further attempts to characterize heuristics.[116] He presents six criteria that can be used to give a more accurate picture of what heuristics are and how they are used in science.[117]

- According to Wimsatt, heuristics are problem solving strategies, but they do not guarantee that the respective problem will be solved or that a research program is called successful.[118]
- They are also cost-effective which makes them especially valuable for the scientific enterprise. Wimsatt points out that, in some cases, where it would be theoretically possible that scientists find a more reliable and accurate way of solving a problem, it is just not affordable.[119] Hence, the need for heuristics in science does not only follow from cognitive constraints, but also from economic and, in the wider sense, socio-political ones.
- Wimsatt also states that heuristics produce errors in a systematic way; he even assumes that those errors are an integral part of the specific design of a heuristic.[120] Hence, using heuristics often allows one to know why certain errors appear and where they come from.
- Heuristics can further be used for the transformation of problems.[121] A good heuristic should thus provide a way to solve the transformed problem and to give rise to the solution of the original one.[122]

---

115 Bechtel and Richardson (2010 {1993}): Discovering Complexity, p. xxvii.
116 Wimsatt (2007): Re-Engineering philosophy for limited beings, Cambridge (Ma), pp. 76-132.
117 See Wimsatt (2007): Re-Engineering philosophy, pp. 76-80. For an intelligible summary of Wimsatt's criteria, see also Lausen (2014): Zur heuristischen Qualität des Reduktionismus, Münster, pp. 91f.
118 Wimsatt (2007): Re-Engineering Philosophy, p. 76.
119 Ibid.
120 Ibid.
121 Ibid., p. 77.
122 Bechtel and Richardson concentrate on this particular function of heuristic strategies, see Bechtel and Richardson (2910 {1993}: Discovering Complexity, Chapter 8 (Reconstituting the phenomenon), pp. 173-192.

- Heuristics do not provide a means to use on any problem; their design specifically matches particular research problems.[123]
- Moreover, heuristics mostly develop out of other, previously existing, heuristics. It thus takes a lot of tuning and previous problem solving strategies to arrive at a powerful heuristic for a given scientific problem.[124]

In short: Heuristics are characterized as (economical) problem-solving strategies which open up new ways of research by systematic error procedures and by the transformation of problems. Furthermore, Wimsatt mentions three heuristic problem-solving strategies in particular, namely, idealization, incompleteness, and (over-)simplification.[125]

Let us now come back to Emil Fischer's usage of the lock-and-key analogy. As mentioned earlier, the question is whether we can speak of the lock-and-key analogy as a heuristic for scientific problem solving. Following Wimsatt's definition of heuristics, I assume that using an analogy heuristically means that it allows for a certain kind of scientific problem-solving by means of the three mentioned strategies, namely idealization, incompleteness and oversimplification.[126] I will now use these strategies to make sense of Fischer's approach in terms of heuristic problem-solving. Fischer's scientific task can be reconstructed as follows: He aimed to explain the nature of chemical reactions between enzymes and sugars, and in particular the phenomenon of enzymatic sugar dissolution.[127] In the absence of detailed information on the composition of enzymes, Fischer used his molecular model of the sugars for the elucidation of enzymatic molecular structure. His starting point was the hypothesis that the configuration of a molecule, in the sense of its three-dimensional interior, had a significant influence on the characteristics of a substance, e.g. on its optical activity and its dissolution potential.[128] The second hypothesis referred to the relation-

---

123 Wimsatt (2007): Re-Engineering Philosophy, p. 80.
124 Lausen (2014): Zur heuristischen Qualität des Reduktionismus, p. 91.
125 Wimsatt (2007): Re-Engineering Philosophy, p. 76f.
126 Ibid., p. 104f.
127 Fischer (1894): Einfluss der Configuration auf die Wirkung der Enzyme, p. 2985 and p. 2992. See also the present study, this chapter, pp. 55-62.
128 Fischer (1894): Einfluss der Configuration, p. 2993. See also the present study, this chapter, pp. 57f.

ship between enzymes and sugars and suggested that these must have complementary molecular configurations in order for a chemical reaction to happen. Here, the lock-and-key analogy came into play, as the complementarity of the reacting enzymes and sugars was described in terms of the fitting between a lock and a key.[129]

As for the question of why a heuristic approach was needed in the first place, one needs to keep in mind that Fischer only had configuration formulas for sugars, but that he did not possess any information on the configuration of enzymes.[130] Would it have been possible to create such configuration formulas for enzymes? Given what was known about enzymes in late 19th century chemistry, the answer is probably no. One of Fischer's important tools in the construction of sugar formulas was van't Hoff's theory of the asymmetric carbon atom and his rule for predicting the number of isomers of the sugar hexose residing therein.[131] There was no such theory with which it was possible to make comparable experimental predictions in the realm of enzyme chemistry. Moreover, let us not forget that Fischer collected a huge amount of experimental information on the behavior of various sugars and that he was actually able to synthesize them; in fact synthesizing different kinds of sugars even provided lots of the experimental data in the first place.[132] Hence, *abstracting* from the case of sugars by means of a molecular model (a model of the molecular structure of a substance) and transferring the knowledge about sugar configuration to the study of enzymes can be seen as Fischer's first heuristic strategy. The second and third strategies (incompleteness and oversimplification) came into play with the introduction of the lock-and-key analogy. From an explanatory perspective, the description of the stereochemical reaction between enzyme and substrate by means of the lock-and-key analogy was *vague and incomplete* in that the respective agents, enzymes and sugars, were not further character-

---

129 Fischer (1894): Einfluss der Configuration, p. 2992; the present study, this chapter, p. 58.

130 Fischer (1924 {1898}): Bedeutung der Stereochemie für die Physiologie, in: Untersuchungen auf verschiedenen Gebieten, p. 134. See also the present study, this chapter, p. 59.

131 See the present study, this chapter, pp. 40-51 and p. 59.

132 See Fischer (1894): Einfluss der Configuration auf die Wirkung der Enzyme, p. 2986; the present study, this chapter, p. 58.

2. The lock-and-key analogy in Fischer's research on fermentation | 75

ized. Furthermore, the formulation of a lock-and-key-like fit between enzymes and sugars left room for a wide range of interpretations concerning the nature of their relationship; i.e. this hypothesis did not *directly* answer how the fitting was accomplished. That is, it did not answer whether it was sufficient that some parts of the molecules complemented each other or whether the whole molecules had to be 'mirror images' of one another. If one considers Fischer's previous work on the sugars, the most obvious interpretation of a lock-and-key-like fit would be that the configuration formulas and more precisely the respective substituents of both enzyme and substrate had to match if manipulated in a certain way.[133] This background, however, was not communicated in the texts in which the lock-and-key analogy was introduced.[134] In some passages Fischer cited his results from the sugar case and he even made use of some of the sugar formulas, but he did not explain any of the strategies involved in manipulating and interpreting these formulas.[135]

The third heuristic strategy which is closely related to the second one and which could also be realized by means of the lock-and-key analogy is oversimplification. The lock-and-key analogy simplified the state of the art in sugar stereochemistry and thereby pointed to its most important findings for pursuing the study of enzymes: 1) That the arrangement of molecules had an influence on chemical and biological features of substances in the

---

133 I have examined the role of paper tools and their manipulation for Fischer's line of reasoning in this chapter, pp. 47-51.
134 See Fischer und Thierfelder (1894): Verhalten der verschiedenen Zucker gegen reine Hefen, in: Berichte (27), pp. 2031-2037; Fischer (1894): Einfluss der Configuration auf die Wirkung der Enzyme, in: Berichte (27), p. 2985-2993; Id. (1895): Einfluss der Configuration auf die Wirkung der Enzyme II, in: Berichte (28); Id. (1898): Bedeutung der Stereochemie für die Physiologie, in: Berichte (26), pp. 60-87; Id. (1909 {1899}): Über die Spaltung racemischer Verbindungen in die activen Componenten, in: Untersuchungen über Kohlenhydrate und Fermente, pp. 890-892.
135 One could of course argue that an explanation like this was not necessary if the formulas had already reached an iconic status at that time. However, especially if one considers the audience for which the respective texts were written – most of them were talks, presented in front of the "Gesellschaft für Naturwissenschaftler und Ärzte" – this is rather unlikely.

first place, 2), that there was something like molecular complementarity between different substances, and 3) that it would be fruitful and promising to concentrate on the stereochemical basis of enzyme-substrate reactions. These hypotheses or considerations were accepted and taken for granted in the field of carbohydrate stereochemistry. In physiology as well as in the emerging field of biochemistry, however, these ideas were still new and exciting in the late 19th century.[136]

Despite its heuristic fruitfulness for the elucidation of the fermentation mechanism, however, the transferability of the concept of molecular geometry by means of the lock-and-key analogy also had its limits. As previously explicated, there is no evidence that Fischer used the analogy in his project on protein synthesis. On the basis of these findings, I have proposed the hypothesis that, in Fischer's eyes, his program on the nature and production of proteins was not sufficiently developed to speculate on the grounds of the analogy. However, as will be shown in chapter 4, others used the lock-and-key analogy to elucidate the nature of proteins. Nonetheless, Fischer's usage of the analogy in fermentation chemistry lasted in the heads of his successors and provided a first orientation for the biochemical study of life.

---

136   Morange (1998): A History, p. 13ff.

3 The making of the lock-and-key model of the antibody-antigen relationship, 1886-1930

In the previous chapter(s), I sketched the origins of lock-and-key analogy usage in biochemistry and explicated its heuristic role in Emil Fischer's program on sugar fermentation in the late 19th century. In what follows, I will have a look at how the lock-and-key analogy and concepts of molecular geometry were used in the realm of immunology. As was shown in the last chapter, Fischer introduced the analogy in order to express the possibility that carbohydrates, which were equal with respect to their chemical structure but had different spatial atomic arrangements, could have distinct effects on microorganisms when fed to them. Fischer's model-projections, chemical formulas that represented the detailed configuration of the carbohydrate molecules, played a central role in the attempt to prove that the molecular geometry of a substance strongly influences its chemical and biological behavior. It was on the basis of these formulas, which hitherto served as construction plans for the synthesis of new and complex carbohydrates, that Fischer made his claim about the causal relationship between the spatial configuration of a substance and its biological effects, and it is likely that not only Fischer's understanding of molecular geometry, but also his terminology of the lock-and-key-like *fit* of enzyme and substrate was inspired by manipulating these modeling devices.

However, in most subsequent biochemical and biomedical contexts in the early 20th century the assumption that molecules have to match geometrically in order to cause a biological function was interpreted with respect to a rather vague understanding of molecular geometry. This understanding was taken as chemical structure plus a spatial dimension that was either not elaborated further or expressed in quasi-chemical terms and pictures.[1] The representation of the antibody-antigen relationship in Paul Ehrlich's immunological studies and the notion of chemical receptors is probably the most famous example for such a modified stereochemical approach and will be of particular interest in this chapter.

In the following, I will address the question of how exactly Paul Ehrlich used the lock-and-key analogy and whether there are similarities to Fischer's usage of the analogy. As will be further explicated in the course of the chapter, Ehrlich did not use it as a heuristic problem-solving tool for model construction in his immunological research. For this purpose, he used metaphorical terms and pictorial representations that shared crucial aspects with the lock-and-key analogy such as the emphasis on the complementary relationship between molecules. The lock-and-key analogy was used in a second step in order to support, justify, and communicate Ehrlich's scientific achievements. I will show that in the course of this communication and justification process Ehrlich's model of immunological reactions was re-interpreted in terms of the lock-and-key analogy and was turned into the lock-and-key model of the antibody-antigen relationship. This re-interpretation process, which took place in different contexts of science (e.g. in popularizing and industrial contexts, as well as within the community of immunologists and immunochemists), started around 1900 and lasted until the 1940s. As will be shown in chapter 4, it had lasting effects on the way in which both Ehrlich's receptor model and the lock-and-key analogy were used in immunochemistry and applied to the realm of molecular biology from the 1930s to the late 1950s. Based on this observation, I will claim that what happened in the phase of re-interpretation between 1910 and the 1950s was a certain kind of model reconstruction.

---

1   See Moulin and Harshav (1988): Text and context in biology. In pursuit of the chimera, in: Poetics Today, 9 (1), pp. 145-161, here pp. 147f. and 185, Morange (2000): A history of molecular biology, p. 14.

3.1 PAUL EHRLICH'S UNDERSTANDING OF IMMUNOLOGICAL SPECIFICITY AND THE LOCK-AND-KEY ANALOGY

Paul Ehrlich was a German physician and immunologist who lived from 1854 to 1915. He is most known for his side-chain theory of immunity for which he was awarded the Nobel Prize in 1908 and for the development of "Salvarsan", the first chemotherapeutic drug against syphilis.[2] Apart from his focus on problems of medical significance, a great deal of Ehrlich's research aimed to explain the concept of immunological 'specificity' from a chemical perspective.[3] 'Specificity' referred to the ability of some organisms or some parts of the organism to selectively react to external influences, e.g. to viruses, bacteria, and venom. Ehrlich and his mentor Robert Koch, the director of the Berlin Institute for the study of infectious diseases ("Institut für Infektionskrankheiten")[4] from 1891 to 1904, both understood 'specificity' in terms of a "rigid and complete one-to-one relationship, according to which a given organism, constant as to morphology and physiology, causes a given disease."[5] In the course of his immunological studies, Ehrlich worked on a chemical explanation for this phenomenon. According to his view, the ultimate cause of a specific immunological reaction was the

---

2   Note that chemotherapy was at first developed as a cure for infectious diseases. It was not until the 1940s that chemotherapeutics were used for the treatment of certain types of cancer. See Chabner and Roberts, Jr. (2005): Chemotherapy and the war on cancer, pp. 65-72.
3   Mazumdar (1995): Species, pp. 80-85.
4   This institute is today known as the "Robert Koch Institut". Paul Weindling provides a detailed analysis of the institute's organizational structure and its development in the late 19th and early 20th century. (Weindling [1992]: Scientific elites and laboratory organization in Fin-de-Siècle Paris and Berlin: The Pasteur Institute and Robert Koch Institute for Infectious Diseases compared, in: Cunningham and Williams [eds.]: The laboratory revolution in medicine, Cambridge and New York, pp. 170-188).
5   Mazumdar (1995): Species, p. 82.

complementary fitting between antibody and antigen molecules as well as between antibody molecules and the cell.[6]

Though many aspects of Ehrlich's immunological theory were criticized in the first half of the 20th century, the link between biological specificity and complementarity continued to play an important role in immunology and in the application of immunological theories and methods to other biological fields, like genetics, embryology and molecular biology.[7]

The similarity between Fischer's lock-and-key analogy and Ehrlich's complementary view of specificity was noticed by contemporaries and has since also been emphasized by historians. Beyond others, Morange, Travis, Mazumdar, Hüntelmann, and Cambrosio et al. have pointed to the appearance of the lock-and-key analogy in Ehrlich's immunological program and to its influence on his symbols and drawings which have in most contexts been interpreted as variants or successors of the lock-and-key analogy.[8] It is known that Ehrlich used the lock-and-key analogy to articulate his views on monocausal specificity between antibodies and antigens by 1900, in his fa-

---

6 Ehrlich defended this view in many of his articles and talks, the most famous being his Croonian Lecture in front of the London Royal Society (See Ehrlich [1900]: On immunity with special reference to cell life, in: Proceedings of the Royal Society of London, 66, pp. 424-448, here p. 434 and 437). Furthermore, Ehrlich and his assistants conducted several studies to experimentally test and elaborate on the idea that immunity was based on a chemical mechanism between complementary substances. Passages that explicitly deal with this issue can be found in, e.g.: Ehrlich and Morgenroth (1904 {1899}): Zur Theorie der Lysinwirkung, in: Arbeiten zur Immunitätsforschung and Ead. (1904 {1889}): Ueber Haemolysine I-V, in: Arbeiten, pp. 16-105. See as well Travis (2008): Models for biological research, pp. 87-90 and Hüntelmann (2012): Paul Ehrlich, pp. 146-151.

7 See Travis (2008): Models for biological research, p. 94f.

8 See e.g. Cambrosio et al. (1996): Ehrlich's "Beautiful pictures", p. 682f.; Cambrosio et al. (2004): Intertextualité et archi-iconicité: le cas des représentations scientifiques de la réaction antigène-anticorps, in: Études de communication, 27, pp. 2-13, here p. 5-7; Travis (2008): Models for biological research: The theory and practice of Paul Ehrlich, p. 87f.; Morange (1998): A history, p. 13, and Mazumdar (1995): Species and specificity, pp. 195f., p. 229 and p. 236.

mous and often cited Croonian Lecture to the British Royal Society.[9] In this lecture, Ehrlich explained and defended his side-chain theory of immunity. The side-chain theory addressed a range of immunological issues, such as the question of the origins of antibodies in the living organism (how are antibodies produced?), the explanation of normal and pathogenic cellular actions (i.e. in the presence of pathogenic agents), and the chemical forces underlying immunological attacks and immune defense.[10] A crucial part of the theory was Ehrlich's receptor model, which proposed a mutual molecular recognition between the cells of an attacked host, a pathogenic substance, and an intermediate substance, called the immune or antibody. Ehrlich assumed that the cell and the antibody could react to the pathogenic agent due to their chemical relatedness ("chemische Verwandtschaft") and suggested a chemical mechanism as the cause of immunological processes.[11] This chemical mechanism was further characterized in terms of the complementary fitting between the molecules of the immunological reactants (cell, antibody, antigen). In the Croonian Lecture, Ehrlich wrote that the relationship between the reactants "must be specific", in that they "must be adapted to one another, as e.g. male and female screw (Pasteur), or as lock and key (Fischer)."[12] In the course of the lecture, Ehrlich presented a range of diagrams in order to explicate what he meant by this "specific" molecular interaction.[13] These drawings came so close to the lock-and-key analogy that the physician and historian of science Ludwik Fleck even referred to them as "lock-and-key symbols."[14]

---

9 See Ehrlich (1900): On immunity with special reference to cell life (Croonian Lecture), p. 434. See as well Cambrosio et al. (1996), p. 682, Mazumdar (1995), p. 195f.; Travis (2008), p. 94.
10 See Silverstein (2009): History of Immunology, pp. 51-56 and 66ff.
11 Ehrlich (1904): Arbeiten zur Immunitätsforschung, p. VII.
12 Ehrlich (1900): On immunity with special reference to cell life (Croonian Lecture), p. 434.
13 Ibid., p. 437.
14 Ludwik Fleck (1935): Genesis and development of a scientific fact, p. 137.

Figure 7

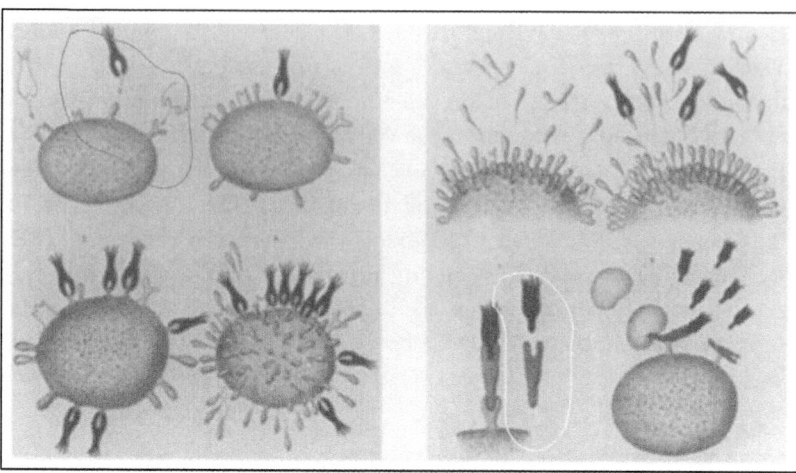

These are two of the six drawings Ehrlich presented in his Croonian Lecture (Ehrlich [1900]). Both drawings ought to show the specific relationship between the antibody or the antigen and the cell molecules (drawing on the left side, emphasis by RM) and the specific relationship between the antigen and the antibody molecules (drawing on the right side, emphasis by RM). Source: Cambrosio, Jacobi and Keating (1996): Ehrlich's "Beautiful Pictures" and the Controversial Beginnings of Immunological Imagery, in: Isis, 84 (4), pp. 662-692, here p. 664.

Beyond the appearance of the lock-and-key analogy in the Croonian lecture and in subsequent publications, the historical literature is not clear about the exact role of the analogy in Ehrlich's research practice. In particular, it is left open whether and how the analogy contributed to Ehrlich's model building processes. There are at least two lines of interpretation: The first one suggests that the analogy inspired Ehrlich's pictorial representations of the antibody-antigen relationship, which, in turn, became a heuristic device for explanation. That is, Cambrosio et al. interpret Ehrlich's famous diagrams as pictorial realizations of the lock-and-key analogy.[15] Yet, these authors also argue for the autonomous heuristic role of the diagrams, suggesting that the lock-and-key analogy would have never been so influential in the early 20th century if these drawings had not made it realizable for a

---

15 Cambrosio et al. (1996): "Beautiful Pictures", p. 689 and Cambrosio et al. (2004): Intertextualité, p. 7.

broad, biologically interested, audience.[16] A second way of interpreting the similarities between Ehrlich's diagrams and the lock-and-key analogy is to assume a common origin; that is to say that the ideas behind these representations were 'in the air' before Fischer used the analogy in his enzymatic research. The historian Pierre Laszlo has called the appearance of Fischer's lock-and-key analogy in organic chemistry and concepts that pointed to a similar mechanism a "co-invention" of similar biochemical concepts and analogies.[17] According to Laszlo, the science of chemistry has "rooted itself on a plethora of hard-to-define, mutually complementary and circular concepts."[18] As will be further explicated in this chapter, Fischer and Ehrlich shared indeed had a common source of influence, namely their training in dyestuff chemistry and their mentor Adolf von Baeyer.[19]

My analysis of Ehrlich's usage of the lock-and-key analogy will show that the analogy itself did not play a major role in Ehrlich's research practice. Ehrlich's 'heuristic toolbox' was, however, filled with concepts, metaphors, and pictures that pointed to similar aspects, such as the stereochemical nature of forces underlying biochemical processes and the importance of molecular complementarity relationships. Yet, the so-called "lock-and-key model for the antibody-antigen-reaction" has been repeatedly ascribed to Ehrlich.[20] This narrative of a linear development from Fischer's usage of the lock-and-key analogy in enzymology to Ehrlich's receptor model is also supported by the historical literature.[21] In contrast to the afore mentioned

---

16 Cambrosio et al. (1996): "Beautiful Pictures", p. 689f.
17 Laszlo (1986): Molecular correlates of biological concepts, in: Comprehensive Biochemistry, 34 A, and Id. (1999): Circulation of concepts, in: Foundations of chemistry, 1 (3), pp. 225-238.
18 Laszlo (1999): Circulation, p. 10.
19 See as well Travis (2008): Models for biological research, p. 85 and Id. (1989): Science as a receptor of technology: Paul Ehrlich and the synthetic dye stuff industry, in: Science in context, 3 (2), pp. 383-408, here p. 388.
20 See Osborn (1937): Complement or Alexin, London, p. 5, and Cambrosio et al. (1996): "Beautiful pictures", p. 683.
21 See Cambrosio et al. (1996): "Beautiful pictures", p. 682f.; Cambrosio et al. (2004): Intertextualité, pp. 2-13, here p. 5-7; Travis (2008): Models for biological research, pp. 88ff.; Morange (1998): A history of molecular biology, p. 13; Mazumdar (1995): Species and specificity, pp. 195f., p. 229 and p. 236; Lenoir

literature, however, I will show that Ehrlich's model of the antibody-antigen reaction was not initially constructed on the grounds of the lock-and-key analogy. Instead, the model was *reconstructed* in terms of the lock-and-key analogy *after* Ehrlich had proposed it. I will argue that this process took place in two stages: Ehrlich's attempt to generalize the receptor model and apply it to the field of infectious chemotherapy from 1906 to 1912 represents the first stage of reconstruction. The second stage is marked by the reception of Ehrlich's scientific achievements between 1900 and the mid-1950s. During this period, Ehrlich's receptor model was vastly re-interpreted as the lock-and-key model of the antibody-antigen reaction. As will be shown, this re-interpretation took place in different scientific contexts, ranging from science popularization to inner-scientific communication (e.g. in articles and speeches), and this had significant consequences for how scientists used the receptor model in the first half of the 20th century.

I will present my claim on receptor-model reconstruction in three steps. First, I will deal with the origins of Ehrlich's receptor model (section 3.2.). Here it will be shown that Ehrlich's views on the chemical nature of immunological processes and his scientific approach were strongly influenced by his training in dye stuff chemistry and his relations to the German chemical industry. Furthermore, it will become clear that Ehrlich worked on the receptor model and on his chemical approach to medical problems before he mentioned the lock-and-key analogy in his immunological articles. Secondly (section 3.3), I will take a closer look at Ehrlich's model-building process and at his 'heuristic toolbox' in order to strengthen the claim that he developed an alternative approach to make use of chemical theories in the realm of immunology, one that was neither based on the lock-and-key analogy nor on Fischer's stereochemical approach in enzymatic chemistry. Instead, Ehrlich made use of concepts and pictures that pointed to the mechanical nature of immunological processes and to the functional criteria of the substances involved in these processes. This alternative approach was inspired by classical theories in structural chemistry (e.g. Kekulé and Pasteur) as well as by methods and concepts developed in the field of industrial

---

(1988): A magic bullet, p. 75; Silverstein (2002): Paul Ehrlich's receptor immunology, p. 83, and Lichtenthaler (1994): 100 Years "Schlüssel-Schloss-Prinzip", p. 2371.

dye stuff chemistry (i.e. Baeyer, Witt), and physiology (i.e. Hoppe-Seyler and Hoffmann).[22] In a third step, I will turn to the process of receptor model reconstruction. I will at first elucidate Ehrlich's usage of the model in his chemotherapeutic research, beginning in 1906 (section 3.4). In this context, the receptor model was used to formulate and justify goals and anticipated results of research. The principle goal of Ehrlich's chemotherapeutic program was the search for chemical "magic bullets"[23], synthetic substances that would only attack the infected cells and leave the healthy ones unharmed. Ehrlich made clear, right from the beginning, that it would take some time to find suitable substances, as this project required a large number of controlled animal experiments.[24] This long search could be justified, however, by Ehrlich's conviction that there was something like a one-to-one specific relationship between the receptors of the cell and the respective chemotherapeutic substance. It will be shown that using the receptor model in the context of the chemotherapeutic program meant to shift the focus from a complex picture of cellular and molecular interactions to the chemical reaction between receptors and chemotherapeutic substances. During this process, the receptor model became autonomous from the side-chain theory of immunity and was reconstructed as a model of the specific reaction between two biologically relevant chemical agents (chemoreceptor and chemotherapeutic agent). In a fourth and last step, I will have a look at the reception of Ehrlich's ideas in inner-scientific as well as in popularizing contexts from the beginning of the 20th century until the mid-1950s (section 3.5). This analysis will uncover that a great deal of the reconstruction process of Ehrlich's receptor model in terms of the lock-and-key analogy took place in the retrospective communication of Ehrlich's scientific achievements.

---

22 Travis (1989): Science as receptor of technology, p. 387 and 390.
23 Hüntelmann (2011): Paul Ehrlich, p. 166f.
24 Ehrlich (1960): Address delivered at the dedication of the Georg-Speyer-Haus (delivered September 6, 1906), in: Himmelweit (Ed.): The Collected Papers of Paul Ehrlich. Chemotherapie, p. 60.

3.2 ORIGINS OF EHRLICH'S RECEPTOR MODEL

As has been pointed out in the historical literature, Ehrlich's receptor model and his side-chain theory were strongly influenced by staining theories and Ehrlich's own physiological work with dyes at the beginning of his career.[25] A main influence in this respect was the work of Adolf von Baeyer, Franz Hofmeister and Felix Hoppe-Seyler, all of which had strong connections to the German dyestuff industry.[26] Von Baeyer, who was the first to synthesize indigo in 1880,[27] trained and supported many "chemical-minded biologists" in his early career which, according to the biochemist and historian Joseph Fruton indicates that he "had an active interest" in the physiological application of chemistry.[28] Hoppe-Seyler's and Hofmeister's chemical research was at its core motivated by physiological questions. Hofmeister received the chair in physiological chemistry in Strasbourg after the death of Hoppe-Seyler who had held the position from 1872 to 1896.[29] Before, Hofmeister had a position at the pharmacological institute in Prague where he was surrounded by scientists who would later become known as early advocates of biochemistry and biomedicine in Europe, such as Karl Spiro and Alexander Ellinger.[30] Both Hoppe-Seyler and Hofmeister were

---

25 Travis (1989): Science as a receptor of technology, p. 385ff. and Dale (1956): Introduction to the Collected Papers, in: Himmelweit (ed.): The Collected Papers of Paul Ehrlich, p. 6.
26 See Fruton (1990): Contrasts in scientific style: Research groups in the chemical and biochemical sciences, Philadelphia. Fruton analyzes and compares the research strategies of five German scientists (Fischer, Baeyer, Hofmeister, Kuhne and Hoppe-Seyler) who, at a certain point in their careers, became known for their important contributions to the establishment of biochemistry. The general thesis of the book is that early approaches in the chemical study of biological phenomena were diverse and to a great extent dependent on the personality of the respective scientist and his style of e.g. laboratory organization.
27 Gerhard Wolf (1970): Die BASF. Vom Werden eines Weltunternehmens, München, p. 22.
28 See Fruton (1990): Contrasts, pp. 125f.
29 Ibid., p. 179f.
30 Ibid., p. 205. As for Hofmeister's scientific orientation, Fruton notes that many scholars who previously dealt with Hofmeister's style of reasoning stated that he

instrumental in the foundation of biochemistry in Germany and especially the former was known for his "interdisciplinary approach", attracting students of various scientific and medical disciplines.[31] The influence of these researchers becomes most visible in Ehrlich's dissertation and habilitation.[32]

3.2.1 Methods and theories of tissue staining

Ehrlich's dissertation was dedicated to new methods of animal tissue staining and the application of synthetic dyes to this end.[33] He extensively studied the chemical literature about the composition of these substances, his major goal being the usage of dye fixing theories for biological and medical purposes.[34] Both Anthony Travis and Timothy Lenoir argue that Ehrlich's early scientific orientation has to be seen in the light of the economic situation of late 19th century Germany, in particular of the fast expansion of German chemical industries, such as BASF, Cassella and Hoechst.[35] As

---

    approached biochemical problems from a biological perspective. Fruton does not explicitly argue against this statement, but demonstrates with a comparative study of Hofmeister's and Fischer's work on proteins that Hofmeister's approach to protein analysis and the attempt to synthesize them was, other than expected, far more chemical then Emil Fischer's, who was known to be an organic chemist through and through (Fruton [1990]: Contrasts, p. 185).

31 See Travis (1989): Science as a receptor of technology, p. 392.

32 See Lenoir (1988): A magic bullet, p. 66 and pp. 79ff.

33 Ehrlich's doctoral thesis "The chemical conception of dyeing" (1878) was dedicated to the staining of bacteria and organs. In his habilitation Ehrlich conducted a follow-up study on the ability of the respective organs (e.g. kidneys, liver) to reduce and safe oxygen from the injected dyes. Dyes were especially useful to observe oxygen reduction by the organs, as in these substances oxygen reduction comes with color loss; thus, if a dye is colorless after being injected into particular parts of the organism, we can conclude that a process of oxygen reduction has occurred (Travis [1989]: Science as a receptor of technology, p. 384).

34 See Travis (1989): Science as a receptor of technology, p. 384.

35 Lenoir (1988): A magic bullet, p. 79. See also Travis (1989): Science as a receptor of technology, pp. 390-393.

global competition between the industrial companies increased, "diversification" in the production of chemicals became more and more important.[36] As a result, the German chemical industry synthesized a wide range of different chemical compounds for which a direct application was not always at hand. Ehrlich's attempt to classify the industrially synthesized dyes according to their biological and medical uses as staining tools thus provided dye-stuff producers with a concrete example for the scientific use of many of these substances.[37] Seen in this context, Ehrlich and many other biomedical researchers filled an open space in the industrial chemical market as their projects involved the application of synthetic chemicals. Hence, whoever was interested in the biological and medical application of chemistry often conducted industry-based research.[38] Unsurprisingly then, research along these lines was marked by a strong collaboration between industrial and scientific institutions in which the industry provided biomedical researchers with a diverse collection of substances. Scientists, on the other hand, continually contributed to industrial growth by giving the chemical industry reasons to proceed in its expansion and diversification.[39] In line with this interpretation, a large part of Ehrlich's writings was devoted to the demonstration of the importance of chemical synthesis for biology, medicine and economic growth. Ehrlich himself noted that the usage of dyes in the realm of toxicology provided a powerful exemplar for the huge potential of structural chemical theories for biomedical innovations.[40]

3.2.2 Chemical constitution in terms of biological function

Another often cited example for the influence of dye stuff chemistry on Ehrlich's biochemical and medical research is his habilitation thesis, "The requirement of the organism for oxygen", in which he addressed the ability

---

36 Travis (1989): Science as a receptor of technology, p. 392.
37 Ibid., p. 390.
38 See Lenoir (1988): A magic bullet; Reinhardt/Travis (2000): Heinrich Caro and the creation of modern chemical industry, and Travis (1989): Science as a receptor of technology.
39 Travis (1989): Science as a receptor of technology, p. 392.
40 Ibid., p. 384.

of the living cell's protoplasm to absorb and store oxygen.[41] Here, Ehrlich did not only use the dyes as stains, but as semi-quantitative measurement devices to test the organs' ability to induce oxygen reactions.[42] Dyes were the ideal candidates for this study as oxygen reduction causes color loss in the case of these substances. Questions that came up throughout the study concerned the identification of the dyes that could enter the cell in the first place and their classification with respect to the conditions under which they were reduced into a leuco product[43] in the respective organs.[44]

Ehrlich was most interested in the relation between the constitution of the respective dye and the reduction of its oxygen molecules by the cell protoplasm. Following Eduard Pflüger's image of the protoplasm as a giant molecule, Ehrlich approached his study from a chemical perspective.[45] The analogy between cells and molecules was based on the assumption that the living cell is driven by the same chemical forces as usual molecules, but

---

41 Ehrlich (1885): Das Sauerstoff-Bedürfniss des Organismus. Eine Farbanalytische Studie, Berlin. For the English translation of the title see Dale: Introduction, in: Himmelweit (1956): The Collected Papers of Paul Ehrlich, London and New York.
42 Ehrlich (1885): Das Sauerstoff-Bedürfniss, p. 19. One could ask at this point if using dyes as staining tools is really different from using them as measurement devices. In order to face this objection, let me briefly explain the difference between Ehrlich's usage of the dyes in his dissertation and habilitation studies. In his dissertation, Ehrlich used dyes to make parts of the cell and single organs visible; providing methods for staining was the major goal. In addition to that, Ehrlich's habilitation involved a quantitative analysis of the chemical distribution within the organism. Here, Ehrlich made use of the dyes' feature of color loss in case of oxygen reduction. The color loss of dyes within particular organs thus gave rise to the respective organ's ability of oxygen take up and storage.
43 In the leuco form dyes are water-soluble and are able to bind cellulosic fibre. (See Christie [2001]: Colour Chemistry, p. 181f.).
44 Ehrlich paid special attention to Alazarinblue, a dye which showed a strong persistency against reduction. It stained only those areas in which the need for oxygen ("Sauerstoffgier") of the cell was particularly high (Ehrlich [1885]: Das Sauerstoff-Bedürfniss, p. 72f.). See as well Travis (1989): Science as a receptor of technology, p. 384f.
45 Ehrlich (1885): Das Sauerstoff-Bedürfniss, p. 8.

that it is of a more complex nature than the hitherto known ones. Ehrlich assumed that every cell activity, e.g. assimilation, growth, and duplication, can be seen as an "expression of a certain chemical organization" and that consequentially, the different functioning of cells is a result of their "specific, peculiar, interior composition."[46] Following this chemical cell model, the question of how exactly molecular constitution and cell function are related became one of the key elements of Ehrlich's investigation. A major challenge of the habilitation thesis thus lay in the further characterization of cell constitution and the causal relationship between constitution and function, in this case the ability to induce oxidation and reduction. Ehrlich's views on molecular constitution were strongly based on Hoppe-Seyler's work on the dye indigo. According to Hoppe-Seyler, functions like light absorption and emission were not caused by the whole molecule, but rather by single atomic groups in a molecule.[47] In line with Hoppe-Seyler, Ehrlich characterized and classified atomic groups as functional units when it came to their constitution ('the atom group that actively absorbs the light').[48] Otto Witt, whose staining theories provided another important resource for Ehrlich's investigation, had a similar understanding of molecular constitution and searched for those atomic arrangements that were "responsible for color."[49]

---

46 The original passage goes as follows: "Es ist nun alles, specifische Lebensleistung, Zeugung, Assimilation, Wachsthum, Vermehrung, Empfindung, Gedanke, Wille-Arbeit der Zellsubstanz (Pflüger) und müssen wir, wenn wir die Leistung als Ausdruck einer bestimmten chemischen Organisation ansehen, folgern, dass die verschiedenen functionirenden Zellen einen specifischen, eigenartigen, inneren Bau besitzen." (Ehrlich [1885]: Das Sauerstoffbedürfniss, p. 8).

47 Ehrlich (1885) quoted Hoppe-Seyler on p. 9: "Wird nun die Lichteinwirkung auf Indigo nicht geändert, wenn es in Indigoschwefelsäure übergeht, so werden wir auch schliessen, dass die Atomgruppe, welche das gelbe Licht lebhaft absorbirt und damit die blaue Farbe des Indigo hervorruft, nicht diejenige ist, an welche die Schwefelsäure sich anfügt, sondern eine andere, welche z.B. bei der Bildung von Indigoweiss getroffen wird (also die Chinogruppe dieses Moleküls)."

48 Ehrlich (1885): Das Sauerstoffbedürfniss, p. 9.

49 Travis (1989): Science as a Receptor of Technology, p. 387. For a more detailed comparison of Witt's staining theories and Ehrlich's early biochemical models, see as well Travis (2008), p. 84.

According to his own statement, applying Hoppe-Seyler's and Witt's considerations to the living protoplasm formed the starting point of Ehrlich's habilitation thesis.[50] In his dissertation Ehrlich had already characterized dyes with respect to their distinct functional components; e.g. the 'color *rendering*' and the 'color *fixing*' components.[51] Following Hoppe-Seyler, Ehrlich now assumed that the cell protoplasm consists of a chemical core of particular structure which is responsible for the cell's activities.[52] Ehrlich claimed that this core at the same time serves as a connection point for a certain class of atomic groups which he called side-chains ("Seitenketten").[53] This passage can be seen as one of the first in which Ehrlich introduced an early version of the side-chain theory, which was later used to explain the origins of antibodies in the context of immune response.[54] Ehrlich did not elaborate on the role of these side-chains in his earlier investigations, but speculated that they are not relevant for specific cell functions ("specifische Zellleistung"), but play an important role for 'life in general', e.g. for physiological processes like fat burning.[55] What he meant by "specifische Zellleistung", however, remains unclear at this point.

50　Ehrlich (1885): Das Sauerstoffbedürfniss, p. 10.
51　Travis (1989): Science as a Receptor of Technology, p. 387.
52　Similar to Fischer, Ehrlich made clear that some features of these atom complexes are more important than others when it comes to their biological effects. The ones he mentions are "Verkettungsart" and "Lagerung" of atoms. (Ehrlich (1885): Das Sauerstoffbedürfniss, p. 8).
53　Ehrlich (1885): Das Sauerstoffbedürfniss, p. 10.
54　After Ehrlich had published his side-chain theory in the context of immunity, he pointed out several times that the study of the protoplasm which he undertook in his habilitation studies formed the starting point for this theory. See e.g. Ehrlich's response to Max von Gruber's attack on the falseness and triviality of the side-chain theory: "Den Ausgangspunkt für die gesamte Betrachtungsweise [by which he means not only the side-chain-theory but as well his chemical and theoretical approach to pharmacological issues, R.M.] bilden die Vorstellungen über das Protoplasma, wie ich sie in meiner Arbeit über das Sauerstoffbedürfnis des Organismus zuerst ausgeführt habe […]." (Gruber Polemik: RAC, 650, Eh 89, 3, and 18).
55　Ehrlich (1885): Das Sauerstoffbedürfnis, p. 10. ("…dass an diesen Kern sich als Seitenketten Atome und Atomkomplexe anlagern, die für die specifische Zell-

Let me shortly summarize the previous section on Ehrlich's usage of dyestuffs in his early career and its significance for the purpose of retracing the origins of the receptor model in immunology. In his dissertation and habilitation theses, Ehrlich approached the question of cellular interactions from a chemical perspective, viewing the cell as a giant molecule that combines with other, less complex, chemical molecules. The means by which he sought to get a better understanding of chemical procedures in the living cell were provided by chemical staining theories and products from the dye stuff industry. In his habilitation thesis, Ehrlich ventured to go one step further and to use dyes not only as a means for cell staining, but also as a foundation, a "chemical model",[56] for his thoughts on cell constitution. For the present purpose, Ehrlich's habilitation allows important insights into his understanding of "chemical constitution", namely the distinctiveness of atomic groupings in a substance with respect to the biological effects they evoke (e.g. light absorbing groups). Hence, although Ehrlich's depiction of the living cell was a chemical one with respect to his focus on interactions between atomic groupings, his understanding of what chemical interactions were, was in fact shaped by the work of biologists and physiologists, e.g. by Hoppe-Seyler and Hofmeister.

Ehrlich made extensive use of existing chemical literature about the structure of dyes and their behavior against other chemical compounds. What is more, every industrial synthesis did not only expand the collection of dyes, but produced at the same time further knowledge about the substances at hand and their chemical relations.[57] Thus, most of the dyes have been elucidated quite well from a structural point of view, as soon as they were available on the industrial market. In most cases, detailed chemical studies of the respective dyes were thus left to others. Nonetheless, Ehrlich did *chemical* work on a theoretical and classificatory basis, using the existing apparatus of structural chemical knowledge about dyes as a classificato-

---

leistung von untergeordneter Dignität sind, nicht aber für das Leben überhaupt. Alles weist darauf hin, dass eben die indifferenten Seitenketten es sind, die den Ausgangspunkt der physiologischen Verbrennung darstellen, indem ein Theil von ihnen die Verbrennung durch Sauerstoffabgabe vermittelt, der andere hierbei consumirt wird.").

56 See Travis (1989): Science as a receptor of technology, p. 386.
57 Ibid.; Lenoir (1988): A magic bullet, p. 78.

ry map for his biomedical investigations. He became an expert in structural chemical theory and due to his collaboration with the dyestuff industry he was able to keep track of and access a wide range of industrially produced products. This in turn helped him to determine quickly which dye was responsible for which effect.

3.3 THE CONSTRUCTION OF THE RECEPTOR MODEL IN THE REALM OF IMMUNOLOGY

In this section I will turn to the construction process of Ehrlich's receptor model in immunology. For this purpose I will take into account two kinds of historical sources, my first source of reference being published articles and books with emphasis on Ehrlich's Croonian Lecture in 1900. The second source I will use comprises the seminal collection of articles, "Gesammelte Arbeiten zur Immunitätsforschung"[58], written by Ehrlich, his closest colleagues and his assistants between 1899 and 1904 and published in 1904 (section 3.3.1). I further considered unpublished material from the Paul Ehrlich Papers, mostly hand- and typewritten notes for his assistants and for himself (section 3.3.2).[59] These notes include short experimental instructions, as well as raw versions of theories and letters, drawings etc. They allow insights into the terminology that Ehrlich and his assistants used to communicate about immunological research tasks and into the usage of symbolic drawings for the same purpose. In turn the communication of Ehrlich and his laboratory assistants and colleagues demonstrates, first, how they approached various studies on the immune system, and secondly, how Ehrlich passed his views on to the next generation of scientists (in this case to his assistants). Together, the considered material shows how Ehrlich and

---

58 Ehrlich (1904): Gesammelte Arbeiten zur Immunitätsforschung, Frankfurt am Main.
59 More specifically, I considered notes from Ehrlich's "Zettelbuch" written in the period from 1900 to 1903 (RAC, RU, 650 Eh 89, Zettelbuch I, II und Carcinom, Box 8, 39, 81, and 82).

the emerging "Ehrlich school"[60] developed a complex system of terms and symbols for the explanation of immunological phenomena.

3.3.1 The side-chain theory of immunity and the receptor model in published articles

A great part of Ehrlich's 'immunological project' was devoted to the explanation of the immune response mechanism in terms of chemical constitution and to the visualization of hidden processes and substances involved therein that could not yet be detected experimentally. Characteristic for Ehrlich's model and at the same time controversial was his postulate of several chemical entities that, by their mechanical interaction, caused the process of immune attack and response, the most basic components being toxine and antitoxine.[61] Ehrlich claimed that each toxine consists of at least two chemical groups, a haptophore group, which has an affinity for a corresponding chemical group of the host's cell, and a toxophore group, which "exercises" the toxic action.[62] In order for the toxine to attack the host's cells, it must consist of a haptophore group; the toxophore group alone would not combine with the cell. One of the central questions for late 19th century immunologists concerned the origin of antitoxines in the living organism and the process of their formation. Ehrlich claimed that antitoxines are produced during normal processes of the cell and then released into the bloodstream of the organism.[63] In the situation of an immune attack these

---

60 For more details on the formation of the "Ehrlich school", see Mazumdar (1995): Species, pp. 82-91 and pp. 381f.
61 Emil Adolf von Behring discovered antitoxine in 1890 in the context of his attempt to find a serum for diphtheria. Behring was one of Ehrlich's closest collaborators during his time at the "Preussische Institut für Infektionskrankheiten" at the Charité in Berlin in the early 1890s.
See http://www.nobelprize.org/nobel_prizes/medicine/laureates/1901/behring-bio.html, 09/27/2018, 20:42.
62 Ehrlich (1900): Croonian Lecture, p. 429.
63 As for the origin of antitoxines and the conception of side-chains, Ehrlich explains that "the antitoxines represent nothing more than side-chains reproduced in excess during regeneration, and therefore pushed off from the protoplasm,

compounds get connected with the cell and function as its chemical extensions, as 'side chains'. The side chains are constituted such that they can as well combine with the toxine and form an "intermediate" between the toxine and the cell.[64] Ehrlich assumed that in order to perform this action side chains had to consist of two haptophore groups, one of which corresponds with the cell and the other with the haptophore group of the toxine. Hence, the side chains are both the preliminary for the immune attack itself, but also a necessary condition for its prevention.[65]

Ehrlich sketched the receptor-model and the side chain theory of immunity most intelligibly in his Croonian Lecture in front of the Royal Society in 1900. The side-chain theory suggested that the cell is composed of so-called side-chains or receptors, expansions of the cell that are chemical in nature and thus capable of connection with (other) molecules.[66] According to the side-chain theory, the receptors are produced in normal metabolic processes and then released into the blood stream, where they eventually function as cell- or antigen-binding (haptophore) groups in the case of an infection.[67] Ehrlich assumed that the host is infected, if molecules of the pathogenic substance bind with the cell's receptors and thereby enter the cell. An infection can be avoided, according to Ehrlich, if the immune body (Immunkörper) connects itself with both the molecules of the receptor and molecules of the pathogenic substance as seen in the center of the image in figure 8. As depicted in the figure below, the immune body was assumed to

---

and so coming to exist in a free state." (Ehrlich (1900): Croonian Lecture, p. 437).

64 Ibid., p. 432.

65 In Ehrlich's understanding the side-chains are important for both the immune reaction and the infection itself: "If the cell of these organs [which are essential for life, R.M.] lacks side-chains fitted to unite with them, the toxophore group cannot become fixed to the cell, which therefore suffers no injury, i.e. the organism is naturally immune." (Ibid, p. 435).

66 Ehrlich/Morgenroth (1899): Zur Theorie der Lysinwirkung, in: Ehrlich (1904): Gesammelte Arbeiten, pp. 1-15, here pp. 13f. "Nach Ehrlichs Definition sind die Seitenketten Träger bestimmter Atomcomplexe, welche befähigt sind, gewisse Atomgruppen an sich zu ketten und so das Molekül des Protoplasmas zu vergrössern."

67 Ehrlich (1900): Croonian Lecture, p. 439.

be capable of intervening in the process of infection, as one of its endings is complementary to the receptor of the cell (c) and the other one to a molecular group of the pathogenic substance (e).

While the first part of the lecture was dedicated to Ehrlich's general model of immunological reaction, illustrated by several examples of bacterial agents and their effect on animal cells,[68] the second part dealt with the characterization of a particular non-bacterial class of substances, the hemolysines, in terms of their chemical constitution. Hemolysines, Ehrlich wrote, are "peculiar toxic bodies" that lead to the destruction of red blood corpuscles, a process which is called hemolysis.[69] Most immunologists before Ehrlich only considered bacteria as toxic agents in immune attack and defense. Ehrlich explained the advantage of using hemolysines over bacteria in immunological research by the fact that the former were, in contrast to most bacteria, easily accessible for test-tube experiments. Hence, the changes that occurred after an injection of hemolysine toxin became much more visible than e.g. in the case of diptheria toxin.[70] Ehrlich ascribed the process of hemolysis to the presence and interaction of two chemical components, a stable one called "intermediate" or "immune body",[71] and a less stable one for which he used the name "complement"; together these components lead to the destruction of red blood cells. The intermediate body was further characterized by two different haptophore groups, one which has a strong chemical affinity for the haptophore group of the red blood cells and one with a weaker affinity for the cells, but which is capable of

---

68 Ibid., p. 429.
69 Ehrlich (1900): Croonian Lecture, p. 442.
70 Ibid.
71 Ehrlich explained that strictly speaking it is problematic to speak of immune bodies in the case of hemolysis, as the respective substances are not always products of a former immunization process. This is why he preferred to call these substances "intermediate bodies" instead of immune bodies. The term "intermediate body" refers to the constitutional and functional characteristics of these substances, but not to their origins in the living organism. (Ehrlich/Morgenroth [1899]: Ueber Haemolysine, in: Ehrlich (1904): Gesammelte Arbeiten der Immunitätsforschung, p. 89).

3. Lock-and-key model construction in immunology | 97

Figure 8

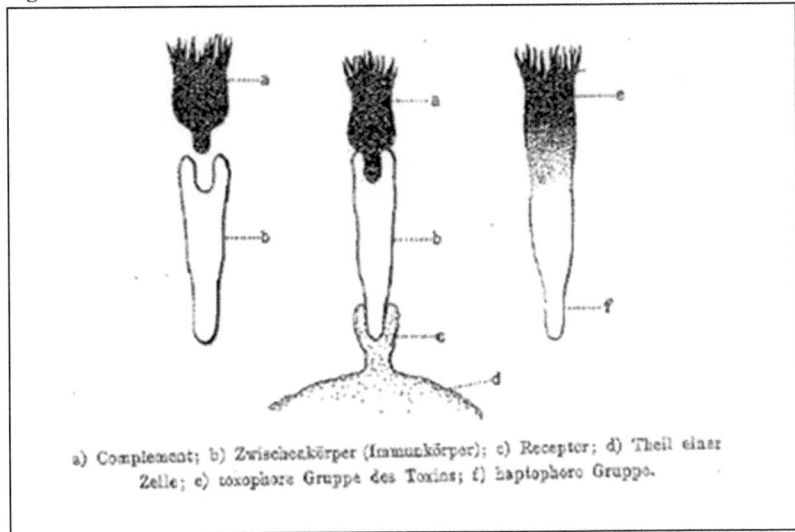

a) Complement; b) Zwischenkörper (Immunkörper); c) Receptor; d) Theil einer Zelle; e) toxophore Gruppe des Toxins; f) haptophore Gruppe.

Ehrlich and Morgenroth (1900): Ueber Hämolysine: Vierte Mitteilung, in: Berliner Klinische Wochenschrift, 37, p. 681.

binding with another component, which Ehrlich stated is part of the serum, i.e. the "Addiment".[72] Ehrlich went on to claim that the experimental results "can only be explained by making certain assumptions regarding the constitution of the intermediate body and the complement".[73] By "constitution" he referred once more to the previously mentioned functional groups. Hence, the intermediate body was further characterized by two haptophore groups, one of which can combine with the "haptophore group of the red blood corpuscles [...] at lower temperature" and "another haptophore group of a lesser chemical temperature, which at a higher temperature becomes united with the 'complement' present in the serum."[74] Ehrlich draws "the conclusion that only with the assistance of the 'intermediate body' or of the 'immune body' can the 'complement,' which leads to the solution [of the red blood corpuscles], become united with the blood corpuscles."[75]

---

72 Ehrlich (1900): Croonian Lecture, p. 444.
73 Ibid., p. 443.
74 Ibid., p. 444.
75 Ibid.

The special case of hemolysis helped Ehrlich to address the "constitution" of toxines and antitoxines in more detail and provided a more complex picture of the proposed mechanism. Apart from the two distinct functional groups, the "haptophore" and the "toxophore" group, Ehrlich introduced a plethora of terms in connection with his experiments on hemolysines, such as "intermediate body", "complement", "amboceptor" and "addiment".[76] Hence, Ehrlich's work on hemolysines is especially interesting with respect to the expansion of his terminological and symbolic network.

I will now have a closer look at Ehrlich's studies on hemolysines and their role in the formulation and further development of the receptor model in terms of chemical constitution. Here, my major source of reference will be the collection of articles "Ueber Haemolysine" (I-VI), written by Ehrlich and his assistant Julius Morgenroth and re-published in "Gesammelte Arbeiten zur Immunitätsforschung".[77] According to various passages in "Ueber Haemolysine" (I-VI), the main theoretical goal of these studies was to prove that there were at least two chemical components responsible for the process of hemolysis.[78] This claim conflicted with Jules Bordet's thesis that hemolysis was caused by the presence of one sensitizing substance called "alexin".[79] Other than Ehrlich, Bordet, a well-received French bacteriologist and immunologists, only used categorizations that pointed to the biological or physiological functions of substances, like e.g. "sensitizing substance", and not to their assumed geometrical shape, divisibility into subparts or their role in a yet unknown chemical mechanism.[80] According to Ehrlich, this controversy was based on conflicting views on the nature of the forces involved in immunological processes, more specifically, on the question of whether these forces ought to be called 'chemical' or not.[81]

---

76 Ibid.
77 Ehrlich (1904): Gesammelte Arbeiten zur Immunitätsforschung, p. 776.
78 Cambrosio et al. (1996): "Beautiful Pictures", p. 668.
79 Ehrlich/Morgenroth (1899): Ueber Haemolysine, 4. Mittheilung, in: Ehrlich (1904): Gesammelte Arbeiten, pp. 86-109, here p. 89.
80 Cambrosio et al. (1996): "Beautiful Pictures", p. 668.
81 Ehrlich/Morgenroth (1899): Ueber Haemolysine, 4. Mittheilung, in: Ehrlich (1904): Gesammelte Arbeiten, pp. 86-109, here p. 89 and p. 123; See as well a note from the "Ehrlich Blöcke" (14. Juni, 1901) in which Ehrlich wrote: "Kampf

Bordet argued that Ehrlich's approach reduced biological phenomena to a purely chemical sphere and even worse to one that seemed to be based on hypothetical considerations about the inner arrangement of molecules.[82] In his view, Ehrlich's images implied the existence of a number of components in the organism which could not be detected experimentally. The sociologists and historians Cambrosio, Jacobi and Keating interpret this controversy as one over the "ontological status of 'antibodies' and related entities."[83] Bordet's major criticism of Ehrlich's model was thus his invention and use of theory-laden terms like e.g. "immune body", "amboceptor", "intermediate body" and "complement". Using these terms and related pictures thus was meant "to postulate the existence of a chemical substance, composed of distinct chemical groups, that would literally insert itself, as a bridge between the cell and the complement, thus allowing the complement to lyse the cell."[84] Despite these controversies, Ehrlich's images and concepts played an important role in the mobility of chemical ideas in biomedical communities, insofar as they made knowledge about the spatial arrangement of atoms intelligible for physicians and biochemists.[85] How far-reaching the integrative function of these pictures was, however, remains an open question, since both chemists and biologists have attacked them for lacking explanatory adequacy. Hence, it is left open in which contexts and by which communities the pictures and symbols were perceived as unproblematic or even insightful and how they could overcome the critics. In this context, the historian Pauline Mazumdar has pointed to the important role of the seminal collection "Gesammelte Arbeiten zur Immunitätsforschung" and the "Zeitschrift für Immunitätsforschung und experimentelle Thera-

---

mit bordet spielt sich auf dem boden der chemischen und achemischen vorstellungen ab." (RAC, RU 650 Eh 89, Zettel-Buch II and Carcinom, Box 8, Folder 2).
82   Cambrosio et al. (1996): "Beautiful Pictures", p. 666f.
83   Cambrosio et al. (1996): "Beautiful Pictures", p. 670.
84   Ibid., p. 668.
85   Ibid., p. 665 and Moulin et al (1988): Text and context in biology, p. 146.

100 | The construction of analogy-based research programs

Figure 9

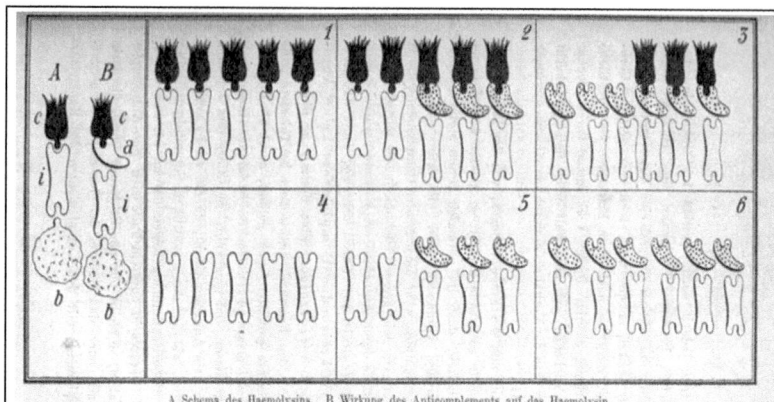

This figure is from Ehrlich's and Morgenroth's article "Ueber Haemolysine, Fünfte Mittheilung", in: Ehrlich (1904): Gesammelte Arbeiten, p. 131.

Figure 10

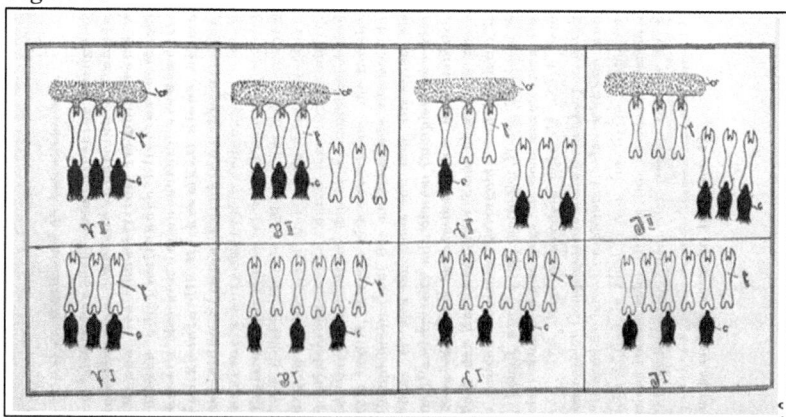

This figure was used by M. Neisser and F. Wechsberg in their paper "Ueber die Wirkungsart der bactericiden Sera", originally published in: Deutsche medizinische Wochenschrift No. 49 (1900) and re-published in "Gesammelte Arbeiten" (1904). The comparison of both figures nicely demonstrates the importance of a particular pictorial style in the tradition of the "Ehrlich school".[86]

pie"[87] in which Ehrlich, his colleagues, and assistants found a medium to assemble and cross-reference their views in support of the receptor model.[88] In part, the influence of Ehrlich's way of thinking and representing can thus be explained by the huge group of upcoming researchers around him – the "Ehrlich school" – and the scientific credibility of the group whose "power structure may be one of the most effective ever formed in science."[89]

As part of the "Gesammelte Arbeiten", the series on hemolysines supported Ehrlich's claim that immunological phenomena were caused by the pairwise interaction of chemical components. Each article in the series summarized a different set of experiments and a theoretical conclusion that picked up the initial question of how the chemical mechanism had to be explained. In the fourth article of the series, Ehrlich and Morgenroth summarized that "in each case that has been considered more carefully, we never found one homogenous body, Buchner's Alexin, but instead the complex hemolysin, consisting of an intermediate body and a complement".[90] They

---

86 On the "persistence" of Ehrlich's ideas and the Ehrlich school in Germany, see Mazumdar (1995): Species, pp. 254-278.
87 This journal was founded by Ehrlich in 1909. Mazumdar notes that the subsequent editors of the journal were – at least until the 1930s - "all representatives of the Koch-Ehrlich-school of immunology." Mazumdar describes the journal as "a house-organ, the voice of the new generation of immunologists, students and grand-students of the bacteriologists of Koch's generation." (Mazumdar (1995), p. 266). See also Sarasin/Berger/Hänseler et al. (2006): Bakteriologie und Moderne. Eine Einleitung, in: Id. (eds.): Bakteriologie und Moderne: Studien zur Biopolitik des Unsichtbaren, pp. 8-43.
88 Mazumdar (1995): Species, p. 257.
89 Ibid., p. 381f.
90 Ehrlich/Morgenroth (1904): Ueber Haemolysine, 4. Mittheilung, Gesammelte Arbeiten, pp. 86-109, here p. 96f. ("Wir haben bei unseren Untersuchungen in allen den Fällen, die wir genauer analysirt haben, nie einen einheitlichen Körper, das Alexin Buchner's, vorgefunden, sondern stets das complexe, aus Zwischenkörper und Komplement bestehende Haemolysin, das in seinen Eigenschaften,

further drew the conclusion that these results "could only be understood from a stereochemical perspective", and that "one has to assume that the complement possesses a haptophore group which reunites with a perfectly matching receptor group of the intermediate body."[91] The consequences of these "observations" are crucial according to Ehrlich and Morgenroth, as they elucidate the relationship between the intermediate body and the complement which can only be described in terms of a "strict specific affinity" (strenge specifische Verwandtschaft).[92] The conducted experiments could thus, according to Ehrlich and Morgenroth, serve as proof for the "duality of the immune body" (Dualität des Immunkörpers) and for the strict specificity of the chemical reactions involved in hemolysis.[93]

However, in the fifth article of the series, Ehrlich and Morgenroth had to modify their claim about the duality of the immune body and the existence of complementary pairs of substances to a certain extent. The reason for this modification was, again, an objection from Bordet who questioned whether the complement and the intermediate body, as well as the residing immune body and the receptor of the cell, have a relationship of strict specificity. Ehrlich's and Morgenroth's response is especially illuminating with respect to the pluralistic interpretation of the receptor concept in (later) con-

---

wie schon betont, vollkommen den immunisatorisch erzeugten Haemolysinen entspricht. Wir werden daher auch annehmen müssen, dass die normalen Heamolysine auch in ihrer Entstehungsweise vollkommen den künstlichen entsprechen.").

91 Ehrlich/Morgenroth (1904): Ueber Haemolysine, 4. Mittheilung, Gesammelte Arbeiten, pp. 86-109, here p. 96f.

92 Ibid., p. 97. "Nach den schon des Näheren erörterten Anschauungen nehmen wir an, dass die haemolytische Wirkung dadurch zu Stande kommt, dass sich Zwischenkörper (Immunkörper) und Complement zu dem komplexen Haemolysin vereinigen. Wir können diese Verhältnisse nur vom stereochemischen Standpunkt aus verstehen und müssen daher annehmen, dass das Complement eine haptophore Gruppe besitzt, die im Zwischenkörper eine genau auf sie passende Receptorgruppe vorfindet. Durch diese Auffassung erhalten aber die Beziehungen zwischen Zwischenkörper und Complement den Charakter strenger Specificität, treten Zwischenkörper und Complement in das Verhältnis strenger specifischer Verwandtschaft."

93 Ibid., p. 109.

texts of neurophysiology.[94] The authors pointed to the "peculiarity of the receptor apparatus" (Eigenthümlichkeiten des Receptorenapparats) of the red blood cells and to the fact that a blood cell probably has "a multiplicity of different types of receptors which suit different immune bodies and hemotoxines".[95] On the one hand this interpretation saved the hypothesis of strict specificity between receptor and complement since it excluded the possibility that there was just one type of receptor for many different kinds of complements. On the other hand, Ehrlich had to shift his prior focus on the dual relationship between chemical compounds to a certain extent, as there were now groups of receptors that combined with groups of complements. Hence, questions concerning the nature of the relationship between two chemical compounds were for the moment lurking in the background, making room for Ehrlich's plurality claims.[96]

---

[94] In neurophysiological contexts the focus shifted from the duality of substances to a more pluralistic interpretation of receptors (multiplicity of receptors in the brain). John Newport Langely was one of the first neurophysiologists who spoke of "receptive substances". See Prüll (2003), p. 334f. and Maehle/Prüll/Halliwell (2002), p. 638f.

[95] Ehrlich/Morgenroth (1904): Ueber Haemolysine, 5. Mittheilung, Gesammelte Arbeiten, pp. 86-109, p. 120f. ("Aus unseren früheren Versuchen über die Isolysine der Ziegen geht hervor, dass wir an einem beliebigen Blutkörperchen eine grosse Zahl verschiedener Typen von Receptoren, die auf differente Immunkörper und Haemotoxine überhaupt passen, anzunehmen haben.").

[96] Ehrlich's focus on the plurality of complements and receptors also becomes visible in other passages in "Ueber Haemolysine" V and VI, for instance in the following quote: "Die Complemente, welche die Aktivirung der normalen und der durch Immunisirung erzeugten Immunkörper (Amboceptoren) vermitteln, besitzen nicht nur für die Immunitätslehre eine hohe theoretische Bedeutung, sondern es dürfte ihnen auch für die normalen Ernährungsvorgänge der Zelle eine wichtige Rolle zukommen. Wir müssen auf Grund der schon früher beschriebenen Versuche annehmen, dass im Blutserum einer bestimmten Thierart nicht nur ein einziges Complement, sondern eine grosse Anzahl verschiedener Complemente existiren. Es ist selbstverständlich, dass nicht alle diese Complemente, die bei einer grossen Reihe verschiedener Species vorkommen, unter sich verschieden sein müssen, sondern es ist als sicher anzunehmen, dass bestimmte Typen eine grosse Verbreitung besitzen, die sich auf mehrere Thierspecies er-

3.3.2 Immunological models in the making: (Quasi-)Chemical terminology, mechanical symbolism and laboratory practice

In this section I will focus on the usage of the previously described network of images and concepts (e.g. 'chemical constitution', 'haptophore' and 'toxophore group', 'complement') in Ehrlich's laboratory practice. My major source of reference is Ehrlich's "Zettel-Buch" and more precisely notes related to the study of hemolysines.[97] These notes were produced between 1900 and 1903 in relation to the previously mentioned series of articles on the nature of hemolysines and their effect on the immune system.[98] The reason for my emphasis on this work is that Ehrlich and his team of assistants have focused on the subject of chemical constitution in their studies on hemolysines and thereby continuously developed and deepened the receptor model. Turning to the Zettel-Buch notes and to the interactions of Ehrlich and his assistants will thus allow insight into the construction and tuning process of the receptor model. Furthermore, the way in which Ehrlich and his assistants interacted with each other via notes and letters exemplifies the pragmatic usage of the previously mentioned concept and images, as these interactions were strongly shaped by the pressure to continuously deliver experimental results and articles.[99] Hence, one can conclude that in these contexts there was less time to strategically garnish a hypothesis with important key words than on more public occasions (e.g. at conferences or public talks); Ehrlich's orders had to be formulated and realized quickly.

The analysis of the Zettel-Buch notes shows that the terminological system which Ehrlich and Morgenroth used in their published articles on hemolysines played a tremendous role in hypothesis formation and in the preparation of first article drafts. One of the first striking observations is that terms and symbols gear into one another. That is, Ehrlich often drew

---

streckt. So erklärt es sich, dass z.B. ein hämolytischer oder bactericider Immunkörper durch die Sera verschiedener Thierarten reactivirt werden kann." (sic.) (Ehrlich/Morgenroth [1904], Ueber Haemolysine, 5. Mittheilung, Gesammelte Arbeiten, p. 124). See as well Mazumdar (1995), p. 380.

97 Zettel Buch II und Carcinom, RAC, Paul Ehrlich Collection, 650 Eh 89, Box 8.
98 See as well Hüntelmann (2011): Paul Ehrlich, p. 238f. and pp. 249ff.
99 See as well Cambrosio et al. (1996): "Beautiful pictures", p. 695.

3. Lock-and-key model construction in immunology | 105

small pictures of combining entities complementing particular terms, the most frequently used ones being: "complement", "anticomplement", "complementophile gruppe", "amboceptor" and "triceptor".[100] As in the published articles, these terms were used to further characterize the reaction between "toxine" and "antitoxine", the two chemical substances present in the case of an attack on particular cells of an animal or human organism.

Furthermore, the analysis of Ehrlich's notes strongly supports the claim that he interpreted experimental results in terms of the (quasi-) chemical terminological and classificatory system mentioned before and that he was confident that his assistants could make sense of his remarks. What is crucial to note is that Ehrlich spoke of "complement", "anticomplement", "amboceptor" etc. as if these were concrete, visible substances that can be treated experimentally; e.g. they can be "heated", "transformed into a more solid bond"[101], "be broken" (by heat)[102], "used to sensitize"[103], "saturate"

---

100 Note to Morgenroth (1901), in: Zettel-Buch II and Carcinom, RAC, Paul Ehrlich Collection, 650 Eh 89, Box 8, Folder 1/3, p. 144; note to Sachs (1902), in: Zettel-Buch II and Carcinom, Box 8, Folder 1/3, p. 55; "Schema für Hemmung" (1902), in: Zettel-Buch III, Box 8, Folder 1/3, p. 60; note to Sachs (1902), Zettel-Buch III, Folder 1/3, p. 86; note to Morgenroth (1901), in: Zettel-Buch III, Folder 1/3, p. 180f.; note to Morgenroth (5. März 1901), in: Zettel-Buch II and Carcinom, Folder 2/3, p. 188; note to Sachs (1901), in: Zettel-Buch II and Carcinom, Folder 2/3, p. 282; note to Morgenroth and Sachs (1. Juli 1901), in: Zettel-Buch II and Carcinom, Folder 2/3, p. 289; "Bildung von Antistoffen" (1901), in: Zettel-Buch II and Carcinom, Folder 1/3, p. 40; note to Morgenroth (1901): Zettel-Buch II and Carcinom, Folder 1/3, p. 53 and p. 55f.

101 Ehrlich to Morgenroth (1902): "Vielleicht sehen Sie einmal nach, ob complement + amboceptor, wenn es etwas höher erwärmt wird – sagen wir 47 ° (es hängt von der Art des complements ab, - event. kann man noch höher gehen, wenn das complement das aushält) in eine festere verbindung tritt und dann reichlicher beides von den rothen blutkörperchen in der kälte gebunden wird." (sic.) (Zettel-Buch II and Carcinom, Box 81, Folder 1/3, p. 54).

102 Ehrlich to Morgenroth (11. April 1902): "Ist einmal untersucht, ob und bei welcher temperatur die anticomplemente caputt gehen?" (sic.) (Zettel-Buch III and Carcinom, Box 82, Folder 1/3, p. 62).

Figure 11

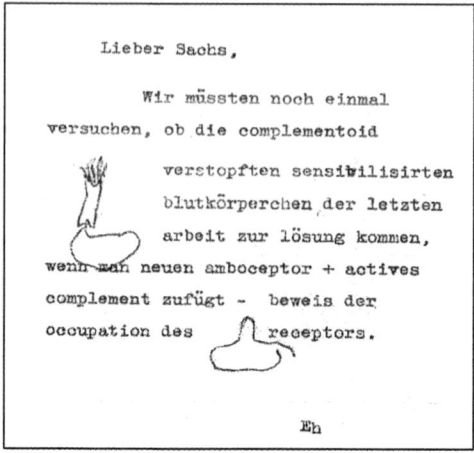

Source: Note from Ehrlich to his assistant and nephew Hans Sachs (1902), in: Zettel-Buch III and Carcinom, RAC, 650 Eh 89, Box 82, Folder 1, p. 86.

and "be saturated"[104] or "dissolved"[105]. This is especially striking in Ehrlich's experimental orders. To give a few more examples: In a letter to Morgenroth, Ehrlich formulated the task that "one would have to find out whether the goat-lysin (ziegenlysin) can be activated by different complements, then, if the reactivation ability of each complement changes, given that several complements are possible and further, if every single complement can be neutralized by its specific anticomplement."[106] In another note to Morgen-

---

103 Ehrlich to Morgenroth (16. Juni 1901): "3) auch mit meerschweinchen-kaninchen amboceptor sensibilisieren" (sic.) (Zettel-Buch III and Carcinom, Box 82, Folder 1/3, p. 272).

104 Ehrlich to Sachs (1901): "Amboceptoren mit 2 complementen sättigen" (Zettel-Buch II and Carcinom, Box 81, Folder 2/3, p. 282).

105 Ehrlich to Sachs (1901): "+ amboceptorauflösung?" (Zettel-Buch II and Carcinom, Box 81, Folder 2/3, p. 282).

106 Ehrlich to Morgenroth (1901): "Dann müsste man feststellen, ob das ziegenlysin (auf ziege geprüft) durch verschiedene Complemente activirbar ist, dann ob falls verschiedene complemente möglich sind, ob das reactivirungsvermögen

Figure 12

```
        Einer thierspecies
ihr eigenes  anticomplement-
serum injiciren. -
        Welcher erfolg?
        Krankheit?
Complementschwund?
Neuerzeugung von  complement.
```

Source: Note from Ehrlich (1900, no exact date), in: Zettel-Buch I, RAC, 650 Eh 89, Box 81, Folder 2/3, p. 252.

roth and Sachs, Ehrlich asked if there have been experiments on the "percentage of complement" (Complementgehalt) in oxen-rabbit serum and notes that it would be "very easy" (kinderleicht) to conduct such an experiment.[107]

Furthermore, the analysis supports the claim that Ehrlich did not use stereochemical concepts or formulas in his investigative immunological practice. Instead, he worked out separate conventions that were both pictorial and textual from the beginning. This finding goes in line with Cambrosio, Keating and Jacoby's thesis that Ehrlich's diagrams played a strong heuristic role in hypothesis formation and model-building.[108] Ehrlich's pictorial approach added something new to the previous approaches to biochemical investigation by establishing a visual link between experimental

---

für jedes einzelne complement wechselt, und weiterhin ob jedes einzelne complement durch sein specifisches anticomplement aufgehoben wird." (sic.) (Zettel-Buch II and Carcinom, Box 81, Folder 1/3, pp.51f.).

107 Ehrlich to Morgenroth and Sachs (1901): "Liegen Erfahrungen über complement gehalt bei unseren Ochsen-Kaninchen Versuchen, ja kinderleicht, wenn man constante mengen hyposensibilisierter blutkörperchen anwendet." (sic.) (Zettel-Buch II and Carcinom, Box 81, Folder 1/3, p. 57).

108 Cambrosio et al. (1996): "Beautiful pictures", p. 689 and Id. (2004): Intertextualité, p. 7.

and conceptual practice. One of the characteristic features of that approach was to leave out the stereochemical details of the substances involved and instead focus on the pictorial representation of the respective chemical interactions. Moulin and Harshaw stress that Ehrlich never actually intended to provide a "precise chemical identification" of the substances he imagined to play a major role in immunological processes.[109] Rather, the chemical terminology was used "to 'make it serious', but any precise definition would involve difficulties".[110] As has been exemplified in this section, the (quasi-)chemical terminology did more than to just provide Ehrlich's approach with the necessary seriousness. In combination with the pictorial representations, Ehrlich's terminology influenced the way in which he approached immunological problems and the internal communication with his assistants.

Moreover, the analysis of the considered notes suggests that Ehrlich did not use the lock-and-key analogy itself as a heuristic tool in his laboratory practice. For this purpose, he used pictures and concepts that were similar to the lock-and-key analogy in that they pointed to similar aspects of the phenomenon in question, such as the complementarity fit between molecules of biological significance and the chemical nature of biological processes. Nonetheless, the lock-and-key analogy supported Ehrlich's immunological ideas in many ways and the aim of the next part of this chapter is to specify how far-reaching this support was. It will be shown that the lock-and-key analogy played an important role both in Ehrlich's career path and

---

109 See Moulin and Harshaw (1988): Text and context in biology, p. 148.

110 Moulin and Harshaw (1988): Text and context in biology, p. 148. The authors analyze the receptor term in various biomedical contexts in the 20th century. What is important to mention is that the authors do not address the lack of precision of the receptor concept as a problem or danger; instead they ascribe an important epistemic and sociological role to loose scientific terminology in that it makes biomedical knowledge accessible to a wide range of scientists and physicians (Ibid. p. 156). With this thesis Moulin and Harshaw anticipate the literature on boundary objects and loose concepts, coined by Griesemer and Leigh Star (1989) and Löwy (1990). However, the authors stick to the communicative role of the receptor concept and do not make particular claims about the terminology's effect on scientific actions, e.g. on the construction of models or theories.

in the subsequent popularity of his biochemical approach. Ehrlich's receptor model was able to gain broad influence due to narratives that highlighted the significance of stereochemical thought and methodology for the exploration of biological phenomena and disease. Using the lock-and-key analogy, these narratives were transported into fields like the chemotherapy of infectious diseases in the late 19th and early 20th century and, as will be examined in chapter 4, to various branches of molecular biology in the second half of the 20th century.

3.4 RECEPTOR MODEL RECONSTRUCTION IN TERMS OF THE LOCK-AND-KEY ANALOGY

Previously, we had a closer look at Ehrlich's receptor model in his immunological work. At first sight, the receptor model seemed to be a lock-and-key model of antibody-antigen relations, that is, an immunological model that is based on the lock-and-key analogy. However, it has been shown that Ehrlich already used some of the concepts and images, which later became known as "lock-and-key symbols",[111] in his habilitation thesis in 1885 – six years before Fischer published his first stereochemical considerations on the sugars and nine years prior to his introduction of the lock-and-key model of enzyme-substrate relations.[112] Thus, there are reasons to doubt that Ehrlich's receptor model is an immediate descendant of Fischer's lock-and-key model of enzyme-substrate relations or that the lock-and-key analogy had a strong heuristic role in this model-building process. However, the

---

111 Fleck referred to Ehrlich's diagrams as "lock-and-key symbols" and wrote that these "became the theory of specificity and for a long time dominated the very depth of the specialized science of serology." (Fleck [1935]: Genesis and development of a scientific fact, p. 137). Cambrosio et al. also cited this passage. Their focus of attention lay on the second part of the sentence in which Fleck designates that Ehrlich's diagrams became the canonical theory of specificity (See Cambrosio et al. [1996]: "Beautiful pictures", p. 670). My emphasis is, however, on the fact that Fleck called these diagrams "lock-and-key symbols" and thereby suggested a strong connection between Ehrlich's drawings and the lock-and-key analogy.

112 See Travis (1989): Science as a receptor of technology, p. 404f.

analogy did play an important role in the continuation and expansion of Ehrlich's research program in the early 20th century. I will argue that in this context the lock-and-key analogy can best be grasped as a tool for model reconstruction.

3.4.1 The receptor model in Ehrlich's chemotherapeutic program

After receiving much attention for his immunological work, Ehrlich turned to the field of infectious chemotherapy[113] in the early beginnings of the 20th century.[114] During this time it became apparent that serum therapy had its limits, as it was not possible, or at least very difficult, to produce antitoxins for every pathogenic agent.[115] At first Ehrlich refrained from applying the receptor model to the study of chemotherapeutics – receptors (substances that were able to bind with antigens and the cell), he thought, only specifically connect themselves with toxins and foodstuffs, but not with synthetically produced drugs.[116] In 1907, in the context of his research on the usage of dyes against trypanosomes[117] and after the neurophysiologist John N.

---

113 Ehrlich coined the term of chemotherapy as a field while investigating the treatment of infectious diseases through the injection of synthesized chemicals. See Bäumler (1989): Farben, Formeln, Forscher, p. 71.
114 See Kasten (1996): Paul Ehrlich. Pathfinder in Cell Biology, in: Biotechnic and Histochemistry, 71 (1), p. 19 and Ulrich Eisenbach (2010): Speyer, in: Neue Deutsche Biographie (NDB), Band 24, Duncker & Humblot, Berlin, p. 674–676. See as well Bäumler (1989): Farben, p. 71.
115 Kasten (1996): Paul Ehrlich. Pathfinder in Cell Biology, p. 19.
116 See Parascandola and Jasensky (1974): Origins of the receptor theory of drug action, in: Bulletin of the History of Medicine, 48, pp. 199-220, here p. 208.
117 Tryponosomes are parasites. Examples for illnesses caused by tryponosomes are sleeping sickness and the American Trypanosomiasis (Chagas'sche Krankheit). Ehrlich tested several synthetically produced dyes against trypanosomes. See Prüll (2003): Part of a scientific master plan? Paul Ehrlich and the origins of his receptor concept, in: Medical History, 47, pp. 332-356, here p. 334. See also Maehle/Prüll/Halliwell (2002): Emergence of drug receptor theory, in: Nature Reviews, 1, pp. 637-641, here p. 639.

Langley had published his ideas on "receptive substances",[118] Ehrlich ventured to speak of chemoreceptors.[119] Soon Ehrlich's experiments with trypanosomes and dyes turned into a research program devoted to the analysis and systematic synthesis of chemotherapeutics against seemingly incurable infectious diseases, the most famous being the drug Salvarsan against Syphilis. It is important to note that there had been similar research in the context of trypanosome therapy before Ehrlich turned to this field. However, these studies were not conducted in the specific realm of chemotherapy, until the necessary organizational resources were available.[120] One important factor in 'the making' of the chemotherapeutic program was the foundation of research institutes that specialized in the treatment of infectious diseases with synthetic chemicals such as the Georg-Speyer wing of the Institute for Experimental Therapy in Frankfurt.[121] In the address delivered at the dedication of the Speyer building, Ehrlich made clear that this particular part of the institute was far more than "one of the laboratories of the ordinary kind"; it represented "something greater than that – the creation of a special type of scientific enterprise which, until now, has had no

---

118 For an analysis of Langley's influence on Ehrlich see Holger-Maehle (2004): Receptive substances, in: Medical History, 48, pp. 153-174, here p. 153.
119 Prüll (2003): Part of a scientific master plan, p. 334f. and Maehle/Prüll/Halliwell (2002): Emergence of drug receptor theory, p. 638f.
120 Ehrlich mentioned himself that Laveran and Thomas as well as Breinl had already tested the atoxyl group against the sleeping sickness in 1903 and in 1905. (Ehrlich [1904]: Gesammelte Arbeiten [III], p. 61f. and p. 90f.).
121 Ehrlich conducted most of his chemotherapeutic research at the Institute for Experimental Therapy in Frankfurt, which was founded in 1896 as the "State Institute for Serum Research and Serum Control" in Berlin-Steglitz. He was announced director of the Institute in its founding year. In 1906, another wing was added to the Institute which was dedicated to the study of chemotherapeutics and named after Georg Speyer, a German banker and philanthropist and the deceased husband of Franzisca Speyer who financially supported the chemotherapeutic wing. After the death of her husband in 1902, Franzisca Speyer made donations to the City hospital in Frankfurt and created the 'Georg and Franzisca Speyer Scholarship Foundation'. See Hüntelmann (2011): Paul Ehrlich, pp. 158-173.

existence in this special form [...] the supreme task of which is to discover specifically therapeutic agents."[122]

Ehrlich paid special attention to the arsenic group, as it seemed to be capable of binding to substances in the body that were analogous to the receptors active in immunological processes. However, he assumed that these chemoreceptors were built in a much simpler way than usual "toxin receptors".[123] Ehrlich deduced this hypothesis from the chemoreceptors' lack of self-organization, their inability to proliferate in the case of chronic intoxication (as in the case of immunization) and from the fact that they were not pushed back into the blood circuit[124] after binding with injected chemi-

---

[122] Ehrlich (1906): Address delivered at the dedication of the Georg-Speyer-Haus, pp. 60f. This is not the only occasion on which Ehrlich pointed to the necessity of a special chemotherapeutic institute for the successful realization of the chemotherapeutic program. For instance, Ehrlich stated in his article about chemotherapeutic trypanosome research: "Who really wants to do chemotherapy will have to be aware of the fact that the discovery of any substance which has an effect against a certain infection is always a question of chance; he also won't expect that he will find the optimal substance right away but he will be content when he finds at all substances with a strong, even if limited, efficacy. The aim of synthetic chemistry is then to improve this initial substance by means of exhaustive synthetic experiments by which new atomic groups will be inserted and others eliminated, so that one arrives at a plurality of chemical compounds from which one can choose the optimal one. I am convinced that for studies like this, there need to be particular institutes for chemotherapy in which the new medications can be synthetically produced and as well tested." (Ehrlich [1960 {1907}]: Chemotherapeutische Trypanosomen Studien, in: Himmelweit [Ed.] [1960]: Paul Ehrlich, Gesammelte Arbeiten, Dritter Band, London/New York/Paris, pp. 81-106, here p. 83f. This study was first published in 1907 in: Berliner Klinische Wochenzeitschrift, 44, pp. 233-236, 280-283, 310-314, and 341-344).

[123] Ehrlich (1909): Chemotherapeutische Trypanosomstudien, in: Beiträge zur experimentellen Pathologie und Chemotherapie, Bd. 1, pp. 97-115.

[124] This goes back to Ehrlich's side-chain theory of immunity. Part of the theory presupposed that the receptors (a synonym for side-chains) fulfill usual functions in the cell metabolism in the absence of an immune attack and that they are pushed back into the blood stream in order to perform these functions after

cals.¹²⁵ Other than in the case of serum-therapy, the goal of chemotherapy was not to cause a defense reaction of the immune system, but to destroy the infectious cells and, in the case of trypanosome chemotherapy, the respective parasite.¹²⁶ Following this model, the main task was the search for specific synthetic substances that were suited to selectively bind particular receptors.¹²⁷ The problem was, however, that the substances, which seemed to be the ideal candidates for this job, were composed of arsenic and, under most conditions, toxic for some of the healthy cells of the body, especially for the nerve cells.¹²⁸ Ehrlich claimed to have found a way to solve the problem of the toxic injury of healthy cells in the so-called "combination therapy" (Kombinationstherapie). Similar to his theory of the constitution of antitoxins, he assumed that an adequate chemotherapeutic would consist of two functional groups, one of which would have an affinity for the pathogenic agent, the "bacteriotrop group", whereas the other one would be ca-

---

    the immune attack has been prevented. See Mazumdar (1995): Species, p. 203f.

125 Ehrlich (1909): Chemotherapeutische Trypanosomstudien, in: Beiträge zur experimentellen Pathologie und Chemotherapie, Bd. 1, pp. 108ff. ("Ich bin nun zu der Anschauung gekommen, daß auch ein Teil der chemischen Substanzen durch Analoga der Rezeptoren aufgenommen wird, die ich im Gegensatz zu den Toxinrezeptoren als C h e m o r e z e p t o r e n [emphasis in original] bezeichnen möchte. Eine besondere Stütze sehe ich in der Tatsache, daß, wie oben bemerkt, der atoxylfeste Stamm auch gegen eine große Reihe von Atoxylderivaten fest ist, die sich chemisch außerordentlich voneinander unterscheiden. Offenbar stellt also hier die Arsengruppe die allen diesen Derivaten gemeinsame Angriffsstelle dar; der Arsenrest wird als solcher gebunden. Man muß sich allerdings die Chemorezeptoren weit einfacher gebaut denken, als die Toxinrezeptoren. Sie besitzen nicht die gleiche Selbständigkeit, können sich dementsprechend bei chronischen Vergiftungen nicht vermehren und werden, da sie sessil sind, auch nicht nach Art der Antikörper in das Blut gestoßen.").

126 Hüntelmann (2011): Paul Ehrlich, p. 166.

127 Ehrlich (1906): Address delivered at the dedication of the Georg-Speyer-Haus (September 6, 1906), in: Himmelweit (ed.) (1960): Paul Ehrlich, Gesammelte Arbeiten, Dritter Band (Chemotherapie), p. 42.

128 Ehrlich (1906): Address Georg-Speyer-Haus, p. 42.

pable of binding with certain organs, the "organitrop group". The goal of the combination therapy was to find the optimal mixture of chemical compounds that consists of the highest possible bacteriotrop group and the lowest possible organitrop group. Ehrlich made clear, right from the beginning, that it would take some time to find suitable substances, as this project required a large number of controlled animal experiments.[129] This long search could be justified, however, by Ehrlich's view that there was something like a specific one-to-one relationship between a pathogenic agents and a certain chemical substance.[130]

The search for chemotherapeutic agents gained first positive results with the synthesis of a number of atoxyl derivatives which seemed to show the kind of specificity Ehrlich was looking for. By 1910 Ehrlich and his colleague Sahachiro Hata were confident that 'dioxy-diamino-arsenobenzol' was the optimal chemotherapeutic substance against a range of infectious diseases and tested it against trypanosomes.[131] Soon, Ehrlich ordered a great amount of 'dioxy-diamino-arsenobenzol' from Hoechst in order to initiate tests against other pathogenic agents in animals. In September 1910, 'dioxy-diamino-arsenobenzol' (then named Salvarsan) was listed as a chemotherapeutic against spirillum diseases in Ehrlich's "compound book" (Präparatebuch). The pathogenic agent that caused syphilis was the most virulent one of the spirilla, and with Salvarsan Ehrlich's group seemed to have found a candidate for its systematic destruction.

The first test series in humans was not conducted by Ehrlich and Hata, but by Johannes Alt from the "Heil- und Pflegeanstalt Uchtspringe" in the beginning of March 1910. After Alt had reported that 0.3 gram of the drug caused a negative Wassermann-reaction and the disappearance of spirochetes in 7 of 23 syphilis patients, J. Hoppe and Ph. Fischer tested Salvarsan more carefully with respect to its physiological effects on the human organism. One of the most important results of this test series was that the arsenic compounds of dioxy-diamino-arsenobenzol seemed to have van-

---

129 Ehrlich (1906): Address Georg-Speyer-Haus, p. 53f.
130 Hüntelmann (2011): Paul Ehrlich, p. 167.
131 The substance was later called "Ehrlich-Hata 606" after the number of substances that had to be tested before Ehrlich and his colleague, the Japanese visiting researcher Sahachiro Hata, discovered the optimal compound. See Hüntelmann (2011): Paul Ehrlich, p. 173f.

ished after 14 days; at least they could not be detected in the blood, urine and stool samples. In the meantime, E. Schreiber, senior physician in the "Altstädtische Krankenhaus Magdeburg", had tested Ehrlich-Hata 606 and reported a positive outcome in 27 cases.[132] Moreover, Ehrlich had asked close colleagues and friends, working in the fields of dermatology and genital diseases, if they would be willing to test the medicament in the clinic. The inquiries from physicians who wanted to test the new wonder drug accumulated after Ehrlich had spoken about the first successes with Salvarsan at the "Kongress für innere Medizin" in April 1910. In October 1910 Ehrlich had already sent over 40.000 samples of the drug to physicians and scientists all over the world and could document more than 260 publications related to test series in clinical contexts. The industrial production of Salvarsan was initiated in the summer of 1910 by the "Farbwerke Höchst" and the final product was available on the commercial market in December.[133] The systematic synthesis of chemicals on a large-scale basis was a necessary condition for the success of the chemotherapeutic program and this, in turn, involved organizational efforts and the inclusion of the industrial companies.[134] The historian Timothy Lenoir argues that the very idea of chemoreceptors was strongly influenced by the dyestuff industry, which developed more and more into a pharmaceutical direction at the beginning of the 20th century.[135] As with Ehrlich's early work on the chemical features of the living protoplasm, his collaboration with dyestuff companies had lasting effects on his research plan and on the size of the chemotherapeutic program.[136]

According to Ehrlich, the real challenge of chemotherapeutic research was the ability of the trained chemist to "aim" (zielen) at certain molecules by means of "chemical variation".[137] A substance with limited side-effects that showed optimal specificity against a pathogenic agents could not immediately be "found"; it had to be *made* "by means of exhaustive synthetic

---

132 Hüntelmann (2011): Paul Ehrlich, p. 193 ff.
133 Ibid, p. 198f.
134 Ibid, p. 168.
135 About the relationship of Ehrlich's industrial connections and the development of his major ideas, see Lenoir (1988): A Magic Bullet, pp. 72-78.
136 Lenoir (1988): A Magic Bullet, p. 75.
137 Hüntelmann (2011): Paul Ehrlich, p. 166.

experiments by which new atomic groups will be inserted and others eliminated, so that one arrives at a *plurality* of chemical compounds from which one can choose the optimal one."[138] Ehrlich demonstrated the possibility of this approach by his determination of the chemical constitution of atoxyl and his subsequent synthesis of arsenic derivatives that, in turn, could be modified such that their toxicity could be regulated.[139] With this first success, the *systematic* influence on all kinds of diseases by the injection of tailor-made chemicals seemed to be within reach.[140]

To summarize, Ehrlich's chemotherapeutic program supported his view of the chemical basis of specificity by realizing it in a context of practical and medical significance. The synthesis of effective drugs against spirillum diseases like Syphilis, Sleeping Sickness, and Malaria on a systematic basis was more or less perceived as a magical act and as the beginning of a new medical revolution.[141] However, the product-oriented character of the chemotherapeutic program also led to a reformulation of goals associated with the model of receptor specificity and, to a certain degree, to a reinterpretation of the model itself. "Specificity" became a feature of a chemical product that could be increased by means of chemical manipulation. The question of whether there really was something like a specific one-to-

---

138 Ehrlich (1960 {1906}: Chemotherapeutische Trypanosomen Studien, in: Himmelweit (Ed.): Paul Ehrlich, Gesammelte Arbeiten, Dritter Band, p. 83f.
139 Ehrlich (1909): Beiträge zur experimentellen Pathologie und Chemotherapie, Bd. 1, Leipzig, pp. 173f. ("Durch die Feststellung der Konstitution war nun die Möglichkeit gegeben, zu einer ungezählt großen Reihe neuer Verbindungen zu gelangen, die alle den Rest einer organisch gebundenen Arsensäure enthalten. Man konnte die verschiedensten Verbindungen in den Ammoniakrest einführen, man konnte sie mit den verschiedensten Säureresten vereinigen, man konnte sie mit Aldehyden kuppeln. [...] Es hat sich hierbei gezeigt, daß je nach den verschiedenen Eingriffen und Umformungen der Arsanilsäure die Verbindung nach Belieben entgiftet oder giftiger wurde.") [Emphasis in original].
140 Hüntelmann (2011): Paul Ehrlich, p. 171-173.
141 Ibid., p. 198 and p. 211.

one relationship between the molecules of a chemical substance and the molecules of a certain pathogenic agent became superfluous, as the medical application of the chemotherapeutic products spoke for itself.[142]

3.4.2 Model Reconstruction in retrospection: The reception of Ehrlich's scientific achievements in Germany and North America

In what follows, I will turn to the reception of Ehrlich's immunological and chemotherapeutic program. It will become clear that Ehrlich's receptor model was transformed into the lock-and-key model of antibody-antigen relations in a gradual process of science communication and re-interpretation. Ehrlich's contemporaries, successors, and the general press reconstructed his scientific path and especially the history of his concept and model building processes in scientific journal articles and reviews, wakes, anniversaries, and in newspaper reports. In terms of the audience that adopted and communicated Ehrlich's work, I broadly distinguish between inner-scientific and external groups and contexts.[143] Ludwik Fleck called attention to

---

142 Ibid., p. 201.
143 It is important to note, though, that one could be more specific here, as there are different kinds of both inner-scientific as well as external groups and contexts. Furthermore, there are intermediate forms of communication that can neither be assigned to internal, nor to external communication, such as knowledge transfer via introductory textbooks. Ludwik Fleck therefore speaks of "intra- and intercollective communication of thought" and of "esoteric" and "exoteric circles" that necessarily overlap at a certain point. (See Fleck [1979 {1935}]: Genesis and development of a scientific fact, p. 111). I did not go into that much detail, as it was more important for the aim of the present study to demonstrate a general tendency of the narratives used in the Ehrlich reception than to report on the more fine-grained differences. A more detailed exploration of the Ehrlich reception would have exceeded the scope of the present study and assumedly deviated from the focus on lock-and-key analogy usage throughout the first half of the 20th century. A follow-up study that concentrates entirely on the forms and contexts of the Ehrlich reception would, however, promise insightful results for the question of how Ehrlich's work was taken up, and more importantly, transformed within these contexts.

the interaction of both inner-scientific and external contexts in the generation of scientific knowledge and claimed that the interpretation of that knowledge by non-scientific communities plays a crucial role in its consolidation and in the development of inner-scientific "thought styles".[144] In particular, the popularization of knowledge has an effect on how that knowledge is passed to the next generation of scientists, or as Fleck put it: The results of popular science will, at a certain point, be "fed back" to the respective scientific communities.[145] Interestingly, Fleck even mentioned Ehrlich's "lock-and-key symbols" as an example of the mutual persuasion of inner-scientific and popular contexts of scientific communication. He stated that

"[t]he achievement of vividness of any knowledge (*eines Wissens*) has a special inherent effect. A pictorial quality is introduced by an expert who wants to render an idea intelligibly to others or for mnemonic reasons. But what was initially a means to an end acquires the significance of a cognitive end. The image prevails over the specific proofs and often returns to the expert in this new role. We can study this phenomenon well by looking at Ehrlich's clear symbolism [....]. The lock-and-key symbols became the theory of specificity and for a long time dominated the depth of the specialized science of serology."[146]

With "lock-and-key symbols" Fleck is clearly referring to the previously mentioned concepts and images in the context of Ehrlich's immunological program (see section 3.3.2), in particular to terms like 'amboceptor', 'complement', and 'anticomplement' and the previously shown diagrams.[147] Fleck's considerations become especially important for the question of how the lock-and-key analogy influenced the community of immunologists in the early 20th century, as they suggest the possibility that science popularization could be one of the crucial sources of that influence. In anticipation of the results of the following analysis, this hypothesis can be confirmed and further specified. It will be shown that the popularization of Ehrlich's scientific achievements in external contexts played a dominant role *in some*

---

144   Fleck (1979 {1935}): Genesis and development, p. 117.
145   Ibid.
146   Ibid.
147   Ibid, p. 64 ff.

periods of reception, while in others it was the community of immunologists and physicians that primarily preserved and shaped the memory of Ehrlich's work. I have identified at least three periods in which Ehrlich's ideas were widely taken up by inner-scientific communities as well as by the broader public in the first half of the 20th century.[148] In addition to the historical literature, I will concentrate on newspaper clippings, speeches

---

148 I have chosen to stop in the 1960s, since the focus of the present study is on the first half of the 20th century. However, results of the historical analysis suggest a fourth period in which Ehrlich's ideas and his biography were widely taken up in Germany. This period can broadly be located in the late 1970s and 1980s. One of the reasons for the increase of the Ehrlich reception during that time is the public appearance of the company Hoechst and their active role in preserving the memory of scientists that influenced the history of the company. In memory of Ehrlich's 125th anniversary, Hoechst initiated a public relation campaign in honor of Ehrlich and his work. In the course of this campaign, the company organized the exhibition "Paul Ehrlich – Forscher für das Leben" and a subsequent publication of the exhibition material (Hoechst, Abteilung für Öffentlichkeitsarbeit (1980): Paul Ehrlich: Forscher für das Leben. Reden zum 125. Geburtstag des Forschers). Apart from these events, the science writer Ernst Bäumler, who worked for Hoechst's public relation department, published several books in the late 1970s and 80s that focused on Ehrlich's person; see e.g. Ernst Bäumler: "Auf der Suche nach der Zauberkugel – Vom grossen Abenteuer der modernen Arzneimittelforschung" (1971), "Paul Ehrlich – Forscher für das Leben" (1979), "Die Rotfabriker" (1988), "Farben, Formeln, Forscher" (1989). Furthermore, Hoechst broadcasted a TV-movie based on Bäumler's book "Auf der Suche nach der Zauberkugel" in 1980. There were other industry-related campaigns around Ehrlich's person as well. For instance, the council of the Central Bank of Germany decided to print a new series of banknotes dedicated to influential German figures (writers, scientists, artists etc.) in 1989. In a row with i.a. Bettina von Arnim, Carl Friedrich Gauß and the Grimm brothers, Ehrlich's face and a molecular stick-and-ball model of the structure of Salvarsan decorated the 200 DM note. The new banknotes were released in 1990. See Lederer/Parascandola (1998): Screening Syphilis: Dr. Ehrlich's Magic Bullet Meets the Public Health Service, in: Journal of the History of the Medical Allied Sciences, 53 (4), pp. 345-70.

and events in homage to Ehrlich, e.g. wakes, anniversaries, exhibitions and movies, with a focus on Germany and North America.[149]

The first reception period began after Ehrlich had formulated his side-chain theory of immunity and reached its peak when he received the Nobel Prize for his achievements in immunology in 1908. During this phase, Ehrlich's work attracted most of the interest within the German community of immunologists and bacteriologists.[150] Here, the interpretation of Ehrlich's theories and models served to either attack or justify his point of view. Most references to the side-chain theory were made in reaction to the ongoing controversy between the Ehrlich school and its critics, such as Jules Bordet, Max von Gruber, and Swantje Arrhenius.[151] As previously mentioned in section 3.2, some of Ehrlich's opponents accused him of proposing a theory on the grounds of insupportable speculations.[152] In defense of Ehrlich, his supporters linked Ehrlich's receptor model to Fischer's lock-and-key model of enzyme-substrate relations, which had already gained a high reputation in biochemical and biomedical contexts by that time.[153] Examples of scientists who justified Ehrlich's speculative model by referring to the authority of Fischer and his lock-and-key model of enzyme-substrate relations are Heinrich Bechhold, Leonor Michaelis, and Max Neisser.[154] Bechhold stated that the side-chain theory was clearly based on Fischer's lock-and-key model and that "no-one can accuse Emil Fischer of messing

---

149 Most of the sources have been selected from the Paul Ehrlich Collection of the Rockefeller Archive Center. The findings are thus based on the collection's preselection of newspaper clippings and articles and do not capture all there is to the reception of Ehrlich's work in Germany and North America. However, the analysis of the selected material provides a first orientation for the 'trends' of the Ehrlich reception and gives insights into some of the narratives that were spread in the reception periods.
150 Mazumdar (1995): Species, pp. 254-270.
151 Ibid, p. 255.
152 See Cambrosio et al. (1996): "Beautiful Pictures", p. 666f.
153 Mazumdar (1995): Species, p. 256.
154 Ibid.

about with fiction."[155] In particular, the receptor model and Ehrlich's notion of the haptophore group, according to Bechhold, corresponds with Fischer's model, which "explained the working of enzymes and glycosides through their stereochemical form, in that they matched each other like 'lock and key'. This is what is meant by Ehrlich's 'haptophore' groups."[156] The physician and biochemist Leonor Michaelis also referred to Fischer's usage of the lock-and-key analogy in order to support Ehrlich's structural-chemical view of specificity: "As the cause of the specific affinity of toxin and antitoxin we must appeal to purely chemical forces, and we understand these in the sense originally meant by Emil Fischer for fermentations, by his image of 'lock' and 'key'."[157] He then went on to link the lock-and-key terminology with Ehrlich's side-chain-theory: "The 'keys' in the living organism are reaction products of the 'lock' and can only be understood in terms of Ehrlich's side-chain-theory."[158] Apart from the direct references to Fischer and the lock-and-key analogy, there are passages, written by Ehrlich's colleagues that supported the interpretation of Ehrlich's model as a lock-and-key model of the relationship between antibody and antigen. For instance, Ehrlich's assistant Julius Morgenroth pointed out that the concept of antibody-antigen linkage, which was crucial for Ehrlich's program, was based on the analogy between complementary chemical interrelations in immunology and those in other scientific contexts, such as in enzymology and toxicology. In 1905 he stated that

"the concept of chemical binding between toxin and antitoxine, an idea borrowed from organic and particularly stereochemistry, is the actual foundation of the theoretical structure built up by Ehrlich and his school. Upon this concept of the binding

---

155 Heinrich Bechhold (1905): Ungelöste Fragen über den Anteil der Kolloidchemie an der Immunitätsforschung, in: Wiener klinische Wochenschrift, 18, pp. 666f. For the translation see Mazumdar (1995): Species, p. 229.
156 Heinrich Bechhold (1905): Ungelöste Fragen, p. 666f.
157 Leonor Michaelis (1908): Physikalische Chemie der Kolloide, in: Alexander von Korànyi and Paul Friedrich Richter (eds.): Physikalische Chemie und Medizin: Ein Handbuch, Leibzig, pp. 341-453, here p. 452. See also Mazumdar (1995): Species, p. 236.
158 Leonor Michaelis (1908): Physikalische Chemie der Kolloide, p. 452. See also Mazumdar (1995): Species, p. 236.

of antigen and antibody rests, by analogy, the relationship of ferment to antiferment, amboceptor to anti-amboceptor, complement to anti-complement, and finally that of toxin, agglutinin and amboceptor to the receptors on the cell."[159]

The second reception phase started with the industrial production and commercial launch of Salvarsan in 1910 and lasted until 1930, the year in which the 20th anniversary of the drug's discovery was celebrated.[160] Ehrlich's discovery of the chemotherapeutic drug was widely discussed in weekly newspapers as well as in medical journals. In this context, Ehrlich was either portrayed as the man with the *magic drug* or as a charlatan.[161] His chemotherapeutic program attracted public discussion in Germany most intensively between 1910 and 1915.[162] Most of the attacks on Salvarsan referred to the commercialization of the product or to its assumed side effects.[163] Especially with respect to the former, the antisemitic climate in early 20th century Germany had its role in the rash denunciation of the drug and its Jewish inventor.[164] As the historian Axel Hüntelmann notes, the criticism had its own "dynamics" and reached its peak in 1914 when Hoechst, the company which sold Salvarsan, Ehrlich and other physicians were openly accused of misconduct and irresponsible behavior in the application of Salvarsan by the press.[165] In subsequent statements and articles, Ehrlich argued on the basis of the principle of chemical specificity that the produc-

---

159 Mazumdar used and translated this passage in order to strengthen her claim that, despite the attacks by Gruber, Arrhenius, and Bordet, Ehrlich's theory was widely accepted by 1905 and that the Ehrlich school was well aware of their dominant position in the German community of immunologists. See Mazumdar (1995): Species, p. 255. The original quote from Morgenroth can be found in: Morgenroth (1905): Ueber die Wiedergewinnung von Toxin aus seiner Antitoxinverbindung, in: Berliner klinische Wochenzeitschrift, 42, pp. 1550-1554, here p. 1550.
160 Budd (2007): Penicillin. Triumph and Tragedy, Oxford, pp. 15-19.
161 Hüntelmann (2011): Paul Ehrlich, p. 211f.
162 Ibid., p. 211.
163 Ibid., p. 211-213.
164 Ibid., p. 214.
165 Ibid., p. 195.

tion of Salvarsan was absolutely justifiable from a scientific standpoint.[166] Discussions about the justification of the medicament within the community of scientists and physicians turned into debates about the exclusion of certain groups of patients (depending on their medical history and physiological condition), the correct dosage and administration of the drug.[167] The responsibility for fatal accidents was thus to be found in the realm of application, that is, Ehrlich claimed that the ideal of a "chemotherapia specifica" was *in principle* correct, the only problem was its implementation in medical practice.[168]

Despite the damaging press of Salvarsan, 1914 was still a good year for Ehrlich's public image. On his 60th birthday in March 1914, Ehrlich's name and picture appeared on numerous newspaper covers and his life-story was printed in a wide range of public as well as scientific journals.[169] Furthermore, the scientific community honored Ehrlich with a *Festschrift* in which his achievements were listed and commented on by experts in the respective scientific fields.[170]

Between 1930 and 1940 the German reception of Ehrlich's work dropped precipitously.[171] This can be explained by the Jewish background of the Ehrlich family. As the organic chemist Bernhard Witkop mentioned in his lecture commemorating Ehrlich and his grandson, Gunther Schwerin, in 1998, Paul Ehrlich's widow, Hedwig Ehrlich, had to flee into exile at the end of the 1930s and was long before exposed to discrimination after she

---

166 Lecture by Ehrlich about Salvarsan therapy, unknown occasion, November/December 1910. (RAC, Eh 89, Series I, Box 3, Folder 11).
167 Hüntelmann (2011), p. 196.
168 Ibid., p. 195.
169 E.g. ‚Die Frankfurter Nachrichten', ‚Die Illustrierte Frankfurter Woche', ‚Die Deutsche Medizinische Wochenschau', ‚Die Naturwissenschaften' and ‚Die Wiener Freie Presse' printed featured articles on the occasion of Ehrlich's 60th birthday. See Hüntelmann (2011): Paul Ehrlich, p. 206.
170 Appolant et al. (1914): Paul Ehrlich. Darstellung seines wissenschaftlichen Wirkens. Festschrift zum 60. Geburtstag des Forschers, Jena. See also Hüntelmann (2011): Paul Ehrlich, p. 206ff.
171 Witkop (1999): Paul Ehrlich and his magic bullets – revisited, in: Proceedings of the American Philosophical Society, 143 (4), p. 540-557, here p. 540.

had established the Paul Ehrlich Prize in 1929.[172] In North America, however, Ehrlich's work was recognized and valued immensely in the 1930s and 40s. Especially in relation to the developing chemical industry and the creation of public health services, Ehrlich's program was perceived as a milestone in the awaited medical revolution and as a promising sign for the systematic influence of the sciences on social problems.[173] The historian Robert Bud notes that the transformation of the medical sciences, e.g. the development of the sulfa drugs[174] in the 1920s and the discovery of penicillin, were perceived as a direct "outcome of Ehrlich's program."[175] Bud ascribes a special role to radio commercials in the perception of the new 'wonder drugs'. In the mid-1930s one-third of the commercially sponsored announcements in the radio "were related to medicine".[176] Furthermore, the first science novels and scientific biographies appeared on the public market in the mid-1920s; e.g. the Nobel Prize winning novel "Arrowsmith" written by Sinclair Lewis in 1925 and one year later Paul de Kruif's "Microbe Hunters" which specifically dealt with Ehrlich's discovery of Salvarsan.[177] On top of that, the US Public Health Service started a long campaign against syphilis including radio commercials and medical movie projects beginning in 1932.[178] The rise of medical commercials and with it Ehrlich's reception in American public contexts reached its peak in 1940 with the Warner Brothers movie production "Dr. Ehrlich's magic bullets" which was nominated for the Academy Awards (also known as the Oscars).[179]

---

172 Witkop (1999): Paul Ehrlich and his magic bullets, p. 540.
173 Bud (2007): Triumph and Tragedy, p. 15.
174 The sulfa drugs are synthetic chemical compounds that circumvent the reproduction of bacteria. For a detailed political history of the production of the sulfa drugs, see John E. Lesch (2007): The first miracle drug: How the sulfa drugs transformed medicine, Oxford.
175 Bud (2007): Triumph and Tragedy, p. 18f.
176 Ibid.
177 Ibid., p. 16.
178 Lederer/Parascandola (1998): Screening Syphilis, p. 349.
179 William Dieterle (director, 1940): Dr. Ehrlich's Magic Bullet, USA. The public relevance of the movie can be illustrated by the fact that Warner Brothers

The movie tells the story of important events in Ehrlich's life and his most memorable scientific achievements.[180] Ehrlich is depicted as a caring doctor who is deeply interested in scientific and especially chemical issues. On top of his responsibilities in the clinic, he spends much time in his small self-arranged laboratory and experiments with chemical staining techniques. In the first scene in which Ehrlich's view on molecular specificity is presented, he talks to Emil von Behring whom he just met for the first time in the hospital and who would go on to become a close colleague of Ehrlich and be awarded with the Nobel Prize in 1901 for his work on serum therapy. The conversation in the movie is centered around the illustration of Ehrlich's chemical approach and his experimental work. Ehrlich comments on his microscopic slides that "it would seem that the chemical make-up of the nuclei has a special affinity for this dye."[181] After Behring had asked him, what he means by "affinity", Ehrlich continues to explain that "the attraction certain atoms possess for certain other atoms causes them to unite and form compounds." This can be shown most strikingly, following Ehrlich, in the case of the dye *methylin blue,* which has a special affinity for the nervous system. Ehrlich presents his experimental slides to Behring who reacts positively shocked: "The whole nervous system blue!" Ehrlich's following explanation is especially insightful with respect to the movie's illustration of his concept of specificity. He notes that *methylin blue* "combines with the nerves, the nerves and nothing else. […] Consider, all depends on discovering a special dye which has an affinity to the substance one wishes to stain."[182] The scene ends with Behring's appropriation of the staining technique just presented by Ehrlich: "This is very important", Behring concludes, "I mean really important."[183]

Ehrlich's view on specificity as a mono-causal relationship between chemical agents is taken up several times in the movie. In the previously

---

produced the movie in cooperation with the US Public Health Service (Lederer and Parascandola (1998): Screening Syphilis, p. 349).
180 The Rockefeller Archive Center holds typewritten dialogue transcripts of the movie (RAC, Eh 89, Box 104, Folder 18). In my following analysis I will, however, refer to the actual movie sequences.
181 Dr. Ehrlich's magic bullets, time: 11:31.
182 Ibid., time: 11:31-13:14.
183 Ibid., time: 13:05.

mentioned scene with Behring and in a number of other scenes which emphasize the innovative aspect of the staining techniques, Ehrlich explains his thoughts on specificity to other physicians. In these scenes, his understanding of specificity is described in simple chemical terms with recourse to the theory of affinity – as it seems, a theory which none of Ehrlich's colleagues in the clinic had ever heard of before.[184] One of these scenes shows Ehrlich at a microbiology congress, where he comments on Robert Koch's talk on tuberculosis and his discovery of the tuberculobaccillus. Ehrlich informs Koch about his staining method and the various sorts of dyes in his laboratory. The dialogue between the two is depicted as follows:

Ehrlich: "It may be possible that one of these [dyes] has an affinity for the tuberculobaccillus."
Koch: "Heh? Affinity? What is it that you say? Ah, come down front, I can't hear you."
[Ehrlich comes down to him]
Koch: "Did you say affinity; did you?"
Ehrlich: "Yes, affinity, the phenomenon of chemical attraction; the will to combine in nature. Now we must compound a dye which will combine with the chemical substance of a microbe and the microbe will be planely seen for it and nothing else on the slide will be stained."
Koch: "Did you ever?"
Ehrlich: "Ever what, Herr Professor?"
Koch: "Ever stain a tuberculobaccillus and nothing else on the slide?"
Ehrlich: "Well, no as a matter of fact, but I have reason to believe it's possible."

In the context of the representation of Ehrlich's chemotherapeutic research, the specificity concept once more becomes the center of focus; it is now depicted as Ehrlich's main theory upon which all of his subsequent work in the field of experimental therapy rests. The movie's emphasis on the foundational and programmatic importance of the concept comes out most strongly in a scene in which Ehrlich talks about his work on syphilis and his first ideas for the chemotherapeutic program at a dinner party hosted by Franziska Speyer.[185] Note that at this point, the lock-and-key analogy is in-

---

184   Ibid., time: 18:30-20:51.
185   Dr. Ehrlich's magic bullets, time: 1:14:00.

troduced for the first time. Due to the importance of this passage, I quote the dialogue between Ehrlich and Mrs. Speyer in full length:

Franziska Speyer (to Ehrlich): "Are you working on a cure?"
Paul Ehrlich: "No, I am working on a new principle."
Fransizka Speyer: "What is this new principle?"
Paul Ehrlich: "Nobody has ever seen a molecule, but let's imagine that they look like this [Ehrlich draws figures on the tablecloth]. Now if we imagine the molecules composing the microbe to look like this keyhole, we can readily see that any chemical molecule that is going to combine with them must be shaped like a key that will fit into them. Now, this is the basis of my theory of affinity. So, after many years I've discovered that arsenic was the key which fitted the molecules of which the microbe is composed of, but it also fits the molecules of which the brain and the nerves are composed of, now we must shape our key so that it fits only the microbes, but since there are thousands of chemical combinations, it will take thousands of experiments."
Franziska Speyer: "Fascinating! Really fascinating. In fact, the most fascinating thing I've ever heard."[186]

This scene conveys Ehrlich's basic idea of chemotherapeutic affinity and specificity, but also the beginnings of the partnership between Ehrlich and his sponsor, Franziska Speyer. The lock-and-key analogy appears in the protagonist's terminology as well as in his drawings which resemble the pictures Ehrlich actually used for his Croonian Lecture (1900) to a great extent (see figure 8). Even more, what the viewer sees on the tablecloth is interpreted with respect to lock-and-key terminology. While the camera turns to the drawings on the tablecloth, we are told by the protagonist that the figure representing the microbe looks like a "keyhole" and that the one which depicts "the chemical molecule" is to be seen as a "key".[187] Thus, the movie gives us a clear interpretation of the drawings in terms of lock-and-key fitting. Also, note that especially Ehrlich's last sentence suggests that the lock-and-key analogy served as a guiding tool in conducting chemotherapeutic experiments and in particular in the adjustment of arsenic compounds (with the goal that these should leave the healthy nervous cells un-

---

186   Ibid., time: 1:14:00-1:15:31.
187   Dr. Ehrlich's magic bullets, time: 1:14:00-1:15:31.

harmed while attacking the infectious cells). The sentence even contains an experimental instruction which is tied back to the previously introduced lock-and-key terminology: "[N]ow we must shape our key [the arsenic compound, R.M.] so that it fits only the microbes, but since there are thousands of chemical combinations, it will take thousands of experiments."[188] Not only does it summarize the experimental goal of the chemotherapeutic project, it also points to the long-lasting efforts and resources that would be needed to successfully realize the project.

The movie provoked numerous positive reactions from the press and the audience. In a letter to Ehrlich's widow, Hedwig Ehrlich, in March 1940, Edward G. Robinson who played the role of Ehrlich wrote that "the Ehrlich movie was an unusual success." "These days", Robinson concluded, "Paul Ehrlich is a name on everyone's lips in America."[189] The historians Susan E. Lederer and John Parascandola support this view and state that the movie "received glowing reviews" by the New York Times and other popular US newspapers.[190] In Germany it was released in 1946, and then broadcasted and re-dubbed into German for West-German Television in 1965 under the title "Paul Ehrlich – Ein Leben für die Forschung".

The next intensive phase of reception of Ehrlich's person and work in Germany and in the US can be located in the mid-1950s, beginning with Ehrlich's 100th birthday in 1954. In Germany, Ehrlich's birthday was celebrated together with Emil von Behring's. On this occasion, a selection of articles in honor of Ehrlich was printed in special issues, medical and scientific journals as well as in weekly newspapers.[191] Furthermore, a post stamp ("Westdeutsche Briefmarke") was created in commemoration of Ehrlich and Behring.[192] In America, Ehrlich's anniversary mobilized even more at-

---

188 Ibid., time: 1:14.
189 Letter from Edward G. Robinson to Hedwig Ehrlich, March 22, 1940, (RAC, Eh 89, Box 63, Folder 18): "Mit Freude kann ich Ihnen mitteilen, dass der Ehrlich Film ein ungewöhnlicher Erfolg bei Presse und Publikum geworden ist. Ueberall wird seine dramatische und menschliche Qualitaet und seine wissenschaftliche Korrektheit geruehmt und der Name Paul Ehrlich ist heute in Amerika in aller Munde."
190 Lederer and Parascandola (1998): Screening Syphilis, p. 355.
191 Hüntelmann (2011): Paul Ehrlich, p. 206ff.
192 Ibid.

tention within the scientific and medical communities. The New York Academy of Medicine organized special events and publications in honor of Ehrlich.[193] The commemoration speeches and articles emphasized the importance of Ehrlich's conceptual achievements for the state of medical and biochemical research in the 1950s. Several speeches and articles told the story of a linear conceptual development from Ehrlich's receptor model to the immunochemical models and theories of their time. According to these interpretations, Ehrlich's concepts and models opened the door for the stereochemical analysis of macromolecules and in many ways anticipated the accomplishments of molecular biology in the mid-20th century. Hugo Bauer, among others, mentioned Ehrlich in a row with Karl Landsteiner, Michael Heidelberger, Felix Haurowitz, and Linus Pauling, all of whom had gained a high reputation for their immunochemical theories in the 1940s and the early 1950s.[194] Hermann Pinkus wrote that, despite the intro-

---

193 See the special issue: Ehrlich Centennial: Annals of the New York Academy of Science (1954), (RAC, 650 Eh 89, Box 63, Folder 6). See as well the invitation letter to the "Twenty-First Series Postgraduate Radio Programme of the New York Academy of Medicine", by the "Committee on Medical Information" (August-October 1954). (Box 63, Folder 21). See also the invitation letter to the "Dinner preceding the Celebration Meeting on Paul Ehrlich's 100th Anniversary at the New York Academy of Medicine, 2 East 103rd Street on Wednesday, March 10, 1954." (Box 63, Folder 21).

194 See Bauer (1954): Paul Ehrlich's influence on chemistry and biochemistry, in: Ehrlich Cent., Annals of the New York Academy of Science, p. 151 (RAC, 650 Eh89, Box 62, Folder 6). Bauer wrote that the "chemical aspects of the formation of antibodies were further developed by Landsteiner, Breinl and Haurowitz, Heidelberger, Pauling, and many others. The chemical character of serological specificity has become more and more manifest. Landsteiner states that antibodies are now definitely recognized as modified globulins. Pauling assumes that antibodies differ from normal serum globulin only in the way in which the two end parts of the globulin polypeptide chain are coiled. The coiling of the end parts is modified under the influence of an antigen and offers a great variety of configurations. These modern concepts appear as refinements and do not differ principally from the old side-chain theory." (Ibid, p. 151). The next chapter will deal more specifically with Landsteiner's, Heidelberger's, Haurowitz's and Pauling's work. Questions of interest concern their

duction of new terms in modern immunology, "we must remember that these modern terms only put a new front on the sound structure that Ehrlich built."[195] This structure, according to Pinkus, was based on "the concept of specific fixation and reaction" and expressed in terms of "specific receptors, toxophor and haptophor groups."[196] In the explanation of Ehrlich's stereochemical approach, Pinkus referred to the "lock-and-key principle" and presented Ehrlich as its founder.

"If we now talk of stereochemical configurations and patterns of electron density, we should remember that the lock-and-key principle and the most important thought of quantitative reactions in immunology are his [Ehrlich's, RM] creations. As a matter of fact, it is in the highly modern branch of quantitative immunochemistry that Ehrlich's ideas find their triumphant realization." [197]

---

understanding of immunological specificity and their references to Ehrlich and the lock-and-key analogy.

195 Hermann Pinkus M.D., Wayne University College of Medicine, Detroit, Michigan: In Commemoration of the 100th Anniversary of the Birth of Paul Ehrlich, Reprinted from the American Journal of Clinical Pathology, Vol. 24, No. 7, July 1954, Printed in U.S.A. Received for publication March 19, 1954, Read at a special clinic for the Alembert Winthrop Brayton Skin and Cancer Foundation in the Department of Dermato-Syphilology at Indianapolis General Hospital, on March 10, 1954, p. 752. (RAC, 650 Eh 89, Box 62, Folder 6).

196 Ibid., p. 752. The complete passage goes as follows: "On the concept of specific fixation and reaction, he [Ehrlich, R.M.] based the whole structure of immunology. [...] He expressed his ideas in the chemical terms of his time in the famous 'side-chain theory' of specific receptors, toxophor and haptophor groups. If this theory is now replaced by 'template' and other concepts, we must remember that these modern terms only put a new front on the sound structure that Ehrlich built. If we now talk of stereochemical configurations and patterns of electron density, we should remember that the lock-and-key principle and the most important thought of quantitative reactions in immunology are his creations. As a matter of fact, it is in the highly modern branch of quantitative immunochemistry that Ehrlich's ideas find their triumphant realization."

197 Pinkus (1954), p. 752.

In a slightly modified reprinted version of this article in 1955, Pinkus mentioned the "lock-and-key principle" once more and explicitly linked it to Ehrlich's receptor model:

"[Ehrlich] theorized that the cell had chemical receptors in the form of side-chains with which it bound food substances coming its way, and that these same receptors would bind foreign substances if they happened to fit the 'haptophore' group according to the lock-and-key principle of Emil Fischer. A receptor thus occupied becomes useless for the cell, which either cannot utilize the foreign substances or is actually damaged if this substance also possesses a 'toxophore' group in some other part of the molecule."[198]

Pinkus was not the only one who described Ehrlich's stereochemical contributions to "modern" immunology and immunochemistry in terms of lock-and-key. E. Witebsky pointed to Ehrlich's usage of the lock-and-key analogy in the illustration of his idea of mono-causal specificity between antigens and antibodies:

"An outstanding characteristic of antibodies lies in their specificity inasmuch as they are directed only against the antigen which caused their production and not against others. Ehrlich liked to use Emil Fischer's famous comparison of the fitting of a key into a keyhole to characterize this specific relationship. The corresponding surface of the two molecules were considered to have complementary shapes, and thus to fit together and provide a stereochemical basis for their union."[199]

Note that Witebsky only cited Fischer, but not Ehrlich when he mentioned Ehrlich's usage of the lock-and-key analogy.[200]

As will be shown in the next chapter, these retrospective interpretations of Ehrlich's scientific achievements in terms of the lock-and-key analogy influenced the way in which the receptor model was understood, and more importantly, the way in which it was used in subsequent immunological re-

---

198 Ibid.
199 Witebsky (1954): Side chain theory, in: Ehrlich Centennial, Annals of the New York Academy of Science, New York. (RAC, 650 Eh89, Box 62, Folder 6, p. 172).
200 Ibid., pp. 168ff.

search programs. This comes out most strongly in the field of immunochemistry in the 1930s and 1940s, where Ehrlich's "lock-and-key model for the antibody-antigen reaction" served as a basis for models of antibody formation.[201] Just as in the context of public reception, in the immunochemical reconstruction in the 1930s and 40s, Ehrlich's receptor model appears to be either a descendant of Fischer's lock-and-key model of enzyme-substrate relations or, even more strikingly, Ehrlich is presented as the founder of the lock-and-key analogy, respectively, the "lock-and-key model of the antibody-antigen reaction".

---

201 Cambrosio et al. (1996): Ehrlich's "Beautiful pictures", p. 683.

4 Lock-and-key foundations for molecular biology: Linus Pauling and the Caltech group, 1930-1960

Previously, we have gained an understanding of the usage of the lock-and-key analogy in the programs of Emil Fischer and Paul Ehrlich at the dawn of the 20th century. I have pointed out two important features of the analogy: The first one referred to its *heuristic role* in the construction of molecular models in fermentation chemistry and enzymology (chapter 2). What made the analogy so tempting in Fischer's program was that it suggested future paths of scientific investigation, but at the same time left much room for context-specific interpretation. The second feature of the lock-and-key analogy which became apparent in the analysis of Ehrlich's program was its *reconstructive role* (chapter 3). Here, the analogy was used for the re-interpretation of Ehrlich's receptor models of immunity and chemotherapeutic reactions in terms of lock-and-key-like relationships. It has been shown that a great deal of this reconstruction process took place in the course of the reception and retrospective communication of Ehrlich's receptor models in the first half of the 20th century.

In this chapter, we will see that the retrospective re-interpretation of Ehrlich's models in terms of the lock-and-key analogy had lasting effects on immunological modeling in the 1920s and 30s and on the integration of different approaches in the developing field of molecular biology in the 1940s and 50s. In early 20th century immunology, the usage of the lock-and-key analogy created a continuum between Ehrlich's view of immunological processes, which was still seen as the foundation of immunochemistry, and new theories of the mechanism of antibody formation, such as the lattice, or framework, theory. The common link, which was constructed between the two generations of immunochemists, was the assumption that molecules from different substances had to complement each other like a lock and a key in order to cause a reaction between antigen and antibody. The immunochemists in the early 20th century ascribed these ideas to Paul Ehrlich, and some of them spoke of Ehrlich's "lock-and-key analogy."[1] In the 1920s and 30s a new theoretical framework was constructed around the lock-and-key analogy, one that to a certain degree shifted the focus from antibodies to antigens and attributed an "instructive" role to the antigen.[2] It was assumed that, at some point of the process, antibody and antigen would have to combine and that this was only possible if they were shaped in a complementary manner. However, new questions were posed in relation to this assumption. Especially the question of how the complementarity of antibody and antigen was possible in the first place became a central issue of immunochemical research. As will be shown, sticking to the lock-and-key analogy brought about a productive shift in the context of early 20th century immunochemistry, and the analogy became heuristically powerful again.

In what follows I will at first sketch the developments in immunochemistry in the 1920s and 30s. The main focus in section 4.1 will lie on the question of how the concepts of immunological specificity and chemical complementarity were linked in the early 20th century. Furthermore, I will pay special attention to the changes that occurred in the theoretical framework of immunochemistry during that time, in particular to the changing role of the antigen in the process of antibody formation. Section 4.2 will then deal with the increasing influence of a new stereochemical interpretation of the concept of complementarity proposed by the chemist Linus Pauling in the late 1930s. Pauling contributed to the development of new physical methods in the study of biochemical phenomena and of modeling techniques and thereby connected new methods of inquiry with the stereochem-

---

[1] See Pauling/Campbell/Pressman (1943): "The Nature of the Forces between Antigen and Antibody and of the Precipitation Reaction", in: Physiological Reviews, Vol. 23. No.3, pp. 203-219, here p. 208.

[2] For a detailed review of the so-called "instruction theories" of antibody formation, see Silverstein (1989): A History of Immunology, San Diego, pp. 66-75.

ical worldview of the late 19th century.³ Furthermore, Pauling's physico-chemical program on the study of antibodies remained to play a significant role in the foundational period of molecular biology in the 1940s and 50s.⁴ As will be shown in section 4.3, Pauling used the lock-and-key analogy in order to extrapolate from his antibody research and to pose the question of how chemical structures were related in terms of molecular fitting and biological function. Pauling's ideas on molecular folding and his template model of antibody formation soon became role models for the study of macromolecules in disciplines other than immunology, e.g. in genetics and embryology.

In section 4.4 I will pay special attention to the usage of the lock-and-key analogy in developing and promoting biochemical research at Caltech. It will be shown that re-interpreting Pauling's template model of antibody formation by means of the lock-and-key analogy led to the integration of different views of biological specificity under the umbrella of one basic model of molecular interactions. Using the lock-and-key analogy in the context of research management set the basis for a universal physico-chemical interpretation of the concept of specificity.⁵ By means of the analogy, a common ground was created between Caltech researchers from different branches of biochemistry (e.g. virology, immunology, bacteriology, embryology, organic chemistry), a precondition for the successful integration of contributions from a wide range of disciplines into one large-scale research program, which I will subsequently call *the Specificity Program*. I use this term in order to emphasize that the projects carried out were programmatic with respect to a common goal, namely, to contribute to and to support (by virtue of their individual studies) a particular view of a common basis of all kinds of specificity reactions. The main idea behind my usage of 'program' in this context is thus that the different projects and corresponding groups involved followed a particular interpretation of the causes

---

3   James (2014): Modeling the Scale of Atoms and Bonds: The Origins of Space-filling Parameters, in: Reinhardt/Klein (eds.): Objects of Chemical Inquiry, Sagamore Beach, pp. 281-320, here p. 283, 309f., and 316.

4   See Kay (1989): Molecular Biology and Pauling's Immunochemistry: A Neglected Dimension, in: History and Philosophy of the Life Sciences, Vol. 11, pp. 211-219, here p. 213.

5   Ibid., p. 214.

of specificity and had an interest in promoting this view within the broader community, e.g. by emphasizing it in their papers and talks.

4.1 SPECIFICITY: IMMUNOCHEMICAL TRENDS AND TRADITIONS

The concept of biological specificity was most influential in the field of immunology where views about the molecular basis of the enzyme-substrate reaction were almost directly applied to the relation between antibody and antigen.[6] Morange traces the concept's origins back to Fischer's usage of the lock-and-key terminology in 1890 and states that it was an "omnipresent" concept in the biological sciences in the first half of the 20th century.[7] Pauline Mazumdar provides a detailed analysis of the specificity concept and its meanings in 20th century immunology, focusing on the dispute between "unitarians" and "pluralists" about the nature of biological phenomena, and the work and career of Karl Landsteiner in particular.[8] Landsteiner, who received the Nobel Prize in 1930 for his early work in blood group genetics,[9] was one of the leading figures in the emerging field of immunochemistry and – as an employee of the Rockefeller Medical Institute – especially well situated in the American community of immunologists and geneticists.[10] Morange describes Landsteiner as an advocate of specificity who argued that "specificity of recognition is an intrinsic property of life" referring to the assumption that organisms are able to produce particular antibodies to all kinds of substances regardless of the complexity of their "chemical nature".[11] Mazumdar distinguishes between two major views of specificity that evolved from the "unitarian" and pluralist tradi-

---

6 Morange (1999): A History, p. 13.
7 Ibid., p. 13. See as well Kay (1989) for the concept's impact in fields of molecular biology. Kay points to the strong tendency of molecular biologists in the 1930s and 40s to concentrate on the specificity of macromolecules (Kay (1989): Pauling's immunochemistry, p. 211).
8 Mazumdar (1995): Species, p. 380.
9 Ibid., p. 317.
10 Morange (1999): A History, p. 14.
11 Ibid., p. 13.

tions and argues that Landsteiner oscillated between these two interpretations.

The pluralistic understanding of specificity was strongly associated with Ehrlich's side-chain theory, respectively the receptor model, and was adopted by members of the influential Koch-Ehrlich group in Berlin.[12] In the tradition of his mentor Robert Koch, Ehrlich defended an absolute view of specificity as complementary one-to-one relationship between molecular units of substances (e.g. antibody and antigen, toxin and antitoxin).[13] Ehrlich's conception of specificity and the resulting side-chain theory was perceived as pluralistic, as it presupposed a large number of chemical groups in the organism, each structurally corresponding to a chemical group in the 'natural world' outside the organism.[14] In opposition to Ehrlich, Max Gruber, Hans Buchner and Alexander Solomon Wiener argued against the idea that specificity was based on a chemical reaction between complementary units and that it could be explained by theories of structural organic chemistry. Instead, their understanding of specificity, which was taken to be the ability of organisms to recognize external influences, was embedded in a unitarian perspective on biological processes.[15] Mazumdar emphasizes that Landsteiner, as a student of Gruber and Fischer, used aspects of both of these specificity concepts, the pluralistic one of the stereochemical Ehrlich-Koch tradition as well as the unitarian one of the emerging community of

---

12 Mazumdar (1995): Species, p. 381f.
13 Ibid., p. 82.
14 Ibid.
15 It remains an open question, however, whether Gruber and others who defended the colloidal view of biochemical reactions were opponents of a particular understanding of specificity or if the colloidal theory as such was incompatible with the concept of specificity, at least in the way in which it was initially conceptualized. In the work of Fischer which is, as mentioned, taken to be the concept's origin, the causes of specific biological processes are clearly located in the realm of organic chemistry and depicted as interrelationships between units of substances. What this episode clearly shows, is that specificity was a vague, if not ambiguous, concept and in many ways a buzz word in the early 20th century biochemistry, especially in the field of immunology. (See Mazumdar (1995): Species, p. 380).

colloid chemists.[16] At first, Mazumdar argues, Landsteiner did not assume well-defined one-to-one relationships between pairs of enzymes and substrates, demarcating specific from non-specific reactions. Instead he conceptualized specificity as "a matter of more or less good fit".[17] Yet in 1924, he started to work with Ehrlich's receptor concept and ventured to postulate absolute specificity for some but not all biochemical systems. Despite this partial adoption of Ehrlich's concept, Landsteiner still remained skeptical concerning the causal nature of interrelations of chemical units and biological phenomena, doubting "that each specificity corresponds to a special compound that could be isolated chemically".[18] As a result, he postulated "two quite separate systems of specificity" in the mid-1920s, "one depending on proteins, with smooth transitions representing his own earlier thinking; the other depending on some other chemical structure and representing Ehrlich's sharp-edged cytotoxin specificity".[19]

In the 1920s and 30s, the two rival perspectives on immunological specificity – the colloidal, gradual interpretation of specificity as a particular stage of a continuous physical transition process between the living and the non-living vs. a one-to-one relationship between molecular units – were approximated to a certain degree. One of the reasons for this development may be found in a more profound understanding of the physical forces operative in the interactions of bio- or macromolecules through the establishment of new measurement methods, especially in the realm of protein research.[20] While most immunochemists still concentrated on the antibody-antigen relationship, the focus shifted from antibodies to antigens as the determining factor in the antibody-antigen reaction. Immunochemists started to attribute an active, even instructive role to the antigen, or as Morange puts it, antibodies were assumed to be "able to interact with antigens because the antigen guided the formation of antibodies."[21] This view was defended most influentially by proponents of the so-called lattice or framework theory, e.g. John Marrack, Friedrich Breinl and Felix Haurowitz, in

---

16 Mazumdar (1995): Species, p. 315.
17 Ibid., p. 284.
18 Ibid., p. 323.
19 Ibid., p. 317.
20 Silverstein (1989): A History, p. 68f.
21 Morange (1999): A History, p. 128.

the early 1930s.[22] The lattice theory suggested that antibodies are formed in a dynamic, at least two-stage process. According to this approach, antibodies are at first normal globulins which are then adjusted to the chemical make-up of the antigenic molecules. This second step, the interaction between antibody and antigen that leads to the final determination of the specific antibody, was assumed to take place on the surface of the antigen.[23] In addition to such conceptual shifts, in 1939, Michael Heidelberger provided a quantitative framework for assessing the mechanism of antibody-antigen combination suggesting that antibody-antigen linkage is a process in which "a given amount of antibody combines with more and more antigen".[24] This picture accounted for a gradual understanding of immunological specificity, as defended by Landsteiner and others. At the same time, the idea that specificity was a result of a chemical one-to-one relationship between complementary molecular units was preserved.[25]

The concept of complementarity was, however, only vaguely defined and the question of how exactly complementarity was involved in the antibody-antigen reaction not explicitly raised, until Linus Pauling entered the discussion in 1940 and combined his general views on the role of shape in macromolecular relationships with the state of the art in the field of immunochemistry.[26] Pauling's "template model" supported a dynamic and gradual understanding of the process of antibody formation as suggested by the so-called lattice or framework theory, but also accounted for the stereochemical traditions of the lock-and-key analogy.[27]

---

22 Cruse/Lewis (eds.) (2010): Atlas of Immunology, Third Edition, p. 74.
23 Silverstein (1989): A History, p. 69.
24 Heidelberger (1939): Quantitative absolute method in the study of antigen-antibody reactions, in: Bacteriological Review, 3 (1), pp. 49-95, here p. 74.
25 See Mazumdar (1989): The template theory of antibody formation and the chemical synthesis of the twenties, in: Mazumdar (ed.): Immunology 1930-1980 – Essays on the history of immunology, Toronto, pp. 13-32.
26 James (2014): Modeling the scale of atoms, p. 308f.
27 Strasser (2006): A World in One Dimension: Linus Pauling, Francis Crick and the Central Dogma of Molecular Biology, in: History and Philosophy of the Life Sciences, Vol. 28, pp. 491-512, here p. 497.

4.2 A NEW STEREOCHEMICAL VIEW OF ANTIBODY-ANTIGEN COMPLEMENTARITY

In 1936, Linus Pauling began his studies on the physico-chemical analysis and synthesis of antibodies at the California Institute of Technology (Caltech).[28] Back then he was known for his work on the linkage between quantum mechanics and organic chemistry, the development of new x-ray diffraction methods and the determination of the chemical structure of hemoglobin.[29] According to his own statement, his interest in immunology began after he met Karl Landsteiner.[30] Mazumdar points out that this encounter substantially influenced both scientists, turning "Landsteiner's thoughts to the significance of bonding in the antigen-antibody reaction, and Pauling's to the problems of antigens and antibodies."[31] Another motivation for Pauling to engage in the question of antibody formation were Felix Breinl and Felix Haurowitz' 1930 paper as well as Stuart Mudd's 1932 article,[32] suggesting that the modification of globulin into a specific antibody was caused by the reordering of the "amino-acid residues in the polypeptide chains".[33] Pauling's major contribution to this new interpretation of the research problem was a detailed interpretation of the reordering process in the globulin chain with the help of new X-ray diffraction methods and the use of modeling devices. As a result, existing concepts of biological specificity

---

28 Kay (1989): Molecular Biology, p. 216.
29 See Paradowsky (2010): "Pauling Chronology". The Ava Helen and Linus Pauling Papers (online), OSU Special Collections & Archives. http://scarc.library.oregonstate.edu/coll/pauling/chronology/page1.html, 10/13/2015, 18:11.
30 Mazumdar (1995): Species, p. 332.
31 Ibid.
32 Breinl and Haurowitz (1930): Untersuchung des Präzipitates aus Hämoglobin und Antihämoglobin-Serum und einige Bemerkungen zur Natur der Antikörper, in: Zeitschrift für Physiologische Chemie, 192, pp. 45-57 and Mudd (1932): A hypothetical mechanism of antibody formation, in: Journal of Immunology, 23, pp. 423-427. See also Morange (2010): What History tells us XX, p. 17.
33 Pauling (1940): A theory of the structure and process of formation of antibodies, in: Journal of the American Chemical Society, 62 (10), pp. 2643-2657, here p. 2643.

were refined and deepened in terms of stereochemical complementarity.[34] Pauling suggested that the changes involved in the chemical make-up of antibodies were more subtle than indicated by the former lattice or framework theory. In his 1940 article "A theory of the structure and process of antibody formation" he argued that a mechanism of antibody-antigen interaction that involves a complete reordering of the amino acid residues (as assumed by Haurowitz and Michael Heidelberger et al.) was very unlikely.[35] In other words, Pauling did not conceive of any reasonable way to model such a mechanism. He therefore suggested a slightly different mechanism of antibody determination, in which "all antibody molecules contain the same polypeptide chains as normal globulin and differ from normal globulin only in the configuration of the chain; that is, in the way that the chain is coiled in the molecule."[36] This mechanism differed from earlier ones with respect to the interpretation and significance of stereochemical configuration, thus opening the door for a detailed study of the spatial structure of normal globulin and antibody molecules. In a subsequent article on "The Nature of the Forces between Antigen and Antibody and of the Precipitation Reaction",[37] Pauling, David Pressman and Dan H. Campbell further explicated the role of configuration in the modification of normal globulin into an antibody:

"If no foreign substance is present, the chain then folds into a stable configuration, characteristic of normal globulin; but if an antigen molecule is present, the chain folds into a configuration stable in the presence of the antigen, that is, into a configuration complimentary to that of a portion of the surface of the antigen molecule."[38]

Pauling's proposal was appealing to immunochemists, in part because it (momentarily) solved a 40 year old puzzle of immunological theories, namely the difficulty to explain the sudden appearance of millions of spe-

---

34 Morange (1999): A History, p. 15.
35 Pauling (1940): A theory of the structure and process, p. 2644.
36 Ibid.
37 Pauling/Campbell/Pressman (1943): The nature of the forces between antigen and antibody and of the precipitation reaction, in: Physiological Reviews, 23 (3), pp. 203-219.
38 Pauling/Campbell/Pressman (1943): The nature of the forces, p. 207.

cific antibodies, 'fitting' to all kinds of invading antigens, including industrially produced ones.[39] In other words, it seemed much more reasonable to explain "the ability of an animal to form antibodies with considerable specificity for an apparently unlimited number of different antigens" by the number of configurations in which one and the same polypeptide chain could be folded, than to assume that normal globulin turns into a structurally completely different substance (the antibody) in the presence of an antigen.[40] The key to the previously mentioned problem was the assumption that small and dynamic changes in the configuration of the globulin chain are the decisive factor in the process of antibody determination.

As highlighted in chapter 2, attempts to study the configuration of organic molecules started in the middle of the 19th century, most famously with Louis Pasteur's studies on tartaric acid and Jacobus Henricus van't Hoff's work on the asymmetric carbon atom. Van't Hoff contributed to the already existent stereochemical movement by proposing a concrete model for the arrangement of atoms in a three-dimensional space. Other than Pasteur and his student Joseph-Achille Le Bel, van't Hoff proposed a molecular model in which all atoms were arranged within a tetrahedron, with the carbon atom at the center.[41] Up to the 1890s, conceptions of molecular geometry were mainly perceived as theoretical constructs. The study of sugars, their classification and synthesis, was one of the fields that contributed to the empirical accessibility of molecular geometry at the end of the 19th century.[42] Especially Emil Fischer's work on sugars and their fermentation was perceived as revolutionary in this respect and became known as one of the first programs that explicitly linked molecular geometry and biological function.[43]

Pauling used the concept of molecular complementarity from late 19th century stereochemistry, the study of the relationship between atoms and their arrangements in the three-dimensional structure of organic molecules, but reinterpreted it with respect to physical principles, such as the electronic

---

39 Silverstein (1989): A History, p. 68f.
40 Pauling/Campbell/Pressman (1943): The nature of the forces, p. 207.
41 See this study, chapter 2, pp. 38ff.
42 Ramsay (1975): Van't Hoff-Le Bel Centennial, p. 74.
43 Ramberg (1995): Chemical Structure, pp. 246ff.; and Mazumdar (1995): Species, p. 194.

van der Waals attraction and electrostatic interaction.[44] He referred to the van der Waals attraction as the "most general force of intermolecular attraction" and considered it to be a vital factor in the antibody-antigen reaction.[45] Yet the reference to electrical interactions did not supply a satisfying explanation for antibody-antigen specificity; such forces could not be the ultimate cause for the specific combination of antibody and antigen. Or to use Pauling's words, the "forces of the van der Waals attraction, hydrogen bond formation, and interaction of electrically charged groups are in themselves not specific; each atom of a molecule attracts every other atom of another molecule by van der Waals attraction."[46] Instead,

"specificity can arise in the interaction of large molecules as a result of the shape of the molecules. [...] A molecule would hence show strong attraction for another molecule which possessed complete complementariness in surface configuration and distribution of active electrically charged and hydrogen-bond forming groups, somewhat weaker attraction for those molecules with approximate but not complete complementariness to it, and only very weak attraction for all other molecules."[47]

The complementarity of the surface configurations of antibody and antigen was thus taken to be a necessary condition for the interaction between the two and for the operation of the mentioned physical attraction forces. Pauling, Campbell and Pressman went even further and suggested a strong causal relationship between the increase and decrease of the van der Waals attraction and the degree of complementarity between the substances involved in the reaction. The van der Waals attraction would be higher, they argued, if the molecules of antigen and antibody came into closer contact, and this in turn depended on the complementary fit of their surface configurations.[48]

---

44 Pauling/Delbrück (1940): The nature of the intermolecular forces operative in biological processes, in: Science (26), pp. 77-79.
45 Pauling/Campbell/Pressman (1943): The nature of forces, p. 205.
46 Ibid., p. 206.
47 Ibid., p.206f.
48 Ibid.

"Hence the electronic van der Waals attraction between two molecules in contact is due practically entirely to interactions of pairs of atoms (in the two molecules) which are themselves in contact; and the magnitude of the attraction is determined by the number of pairs of atoms which can be brought into contact. In consequence, two molecules which can bring large portions of their surfaces into close fitting juxtaposition will in general show much stronger mutual attraction than two molecules with less extensive complementariness of surface topography."[49]

The increase of weak chemical bonds and the van der Waals attraction served as an indicator for the complementary fit of antigen and antibody, "confirming the model proposed half a century earlier by Emil Fischer", as Morange puts it.[50] Yet, in contrast to Fischer and his contemporaries, Pauling and his colleagues were able to predict the shape of a molecule "by locating the atoms within the molecule with use of bond distances and bond angles and then circumscribing about each atom a spherical surface corresponding to its van der Waals radius."[51]

Pauling, Campbell and Pressman emphasized the close connection between the "new complementariness theory", as they called this combination of the lattice and the template theory, and the lock-and-key analogy.[52] The authors claim that their theory of complementarity "is not greatly different from some earlier proposals, such as Ehrlich's lock-and-key analogy, but it differs greatly from others".[53] Apart from the fact that Pauling and his colleagues did not mention Fischer in their reference to the lock-and-key analogy, this passage sheds light on how Ehrlich's side-chain theory was 'inherited' by his successors. Pauling, Campbell and Pressman, as well as Haurowitz, supported what they thought of as Ehrlich's hypothesis about the lock-and-key-like mechanism of antibody-antigen-binding, but they rejected the other parts of the side-chain-theory according to which antibodies

---

49 Pauling/Campbell/Pressman (1943): The nature of forces, p. 205.
50 Ibid.
51 Ibid., p. 204.
52 Ibid., p. 208.
53 Ibid., p. 208. Here, Pauling et al. mainly refer to Buchner who defended the colloidal view of immunological reactions, and to his contemporaries Burnet and Jordan. In the context of this article they speak of "Ehrlich's lock-and-key analogy."

were "formed by the regeneration of receptors preformed in the living organism".[54] According to Haurowitz, this part of Ehrlich's theory had to be rejected because it implied a natural correspondence between the chemical make-up of substances originating in the organism (antibodies) and the make-up of foreign substances entering it (antigens).[55] The hypothesis that specific antibodies were formed by receptors, previously involved in normal processes of the organism, was reasonable as long as their specific antigens were taken to be limited products occurring in the natural world. This was proven to be wrong by Landsteiner and colleagues who showed that modified antigens whose "specificity was altered by the introduction of certain chemical groups into the antigen molecule" also led to the formation of specific antibodies.[56] As previously explained, both the framework and the template theory replaced Ehrlich's hypothesis of preformed receptors by concentrating on the physico-chemical transformation process which turned normal globulin into an antigen-specific substance, the antibody. Hence, the framework and template theory assumed that globulin was itself transformed into the antibody, without the necessity of yet another coupling between the macromolecule and some other substance.

As has been shown in section 4.1, the lattice or framework theory already accounted for the concept of molecular complementarity. Proponents of this theory, such as Haurowitz and Heidelberger, did not specify, however, whether complementarity was a product of the antibody-antigen reaction or a necessary condition for the reaction to happen.[57] By deepening the stereochemical and physical dimension of specificity,[58] Pauling, Campbell and Pressman answered this question in favor of the latter option: In order to account for the causation of "strong mutual attraction", antibody and antigen molecules were conceptualized as complementary in their "surface topography".[59]

---

54 Haurowitz (1947): Antibodies: Their Nature and Formation, in: The Lancet (Special Articles), Jan. 25, p. 149-151, here p. 149. See as well Pauling/Campbell/Pressman (1943): The nature of the forces, p. 208.
55 Haurowitz (1947): Antibodies, p. 149.
56 Ibid.
57 See Silverstein (1989): A History, p. 318.
58 Morange (1998): A History, p. 127.
59 Pauling/Campbell/Pressman (1943): The nature of the forces, p. 206.

4.3 PAULING'S UNIVERSAL MOLECULAR AGENDA: THE IMPORTANCE OF COMPLEMENTARITY FOR BIOCHEMICAL REACTIONS

Pauling's immunochemical work that, among other things, led to the template theory and the attempt to create a construction plan for the synthesis of antibodies, was deeply entangled with the larger research program on chemical and physical features of proteins. Apart from the fact that antibodies were assumed to be proteins since the 1920s,[60] the link between these two research fields was established by the general assumption that two chemical agents must have complementary shapes in order to bring about a reaction of biological importance, one that allows for the specificity of a substance or an organism.[61] From a socio-political perspective, Pauling's program on immunochemistry was also a means to ensure financial support of protein-related research in the war and post-war period. As Kay points out, research on antibodies was of "critical importance" as it provided "continuity, prestige and resources to Caltech's molecular biology program."[62]

It is thus not surprising that many of Pauling's articles and talks in the 1940s and 50s were devoted to the demonstration of the generality of complementary reactions and their major role for biological specificity. He started to emphasize complementarity as a cause of biological specificity *in general* in his joint article with Max Delbrück on "Intermolecular forces operative in biological processes" in 1940.[63] In this article Pauling and Delbrück stated "that complementariness should be given primary consideration in the discussion of the specific attraction between molecules and

---

60   Pauling/Campbell/Pressman (1943): The nature of the forces, p. 206.
61   See e.g. Pauling (1940): A theory of the structure, p. 2643f; Pauling/Delbrück (1940): Intermolecular forces operative in biological processes, in: Science, 92 (2378), pp. 77-79; Pauling/Campbell/Pressman (1943): The nature of the forces, p. 206f; Pauling (1946): "Molecular Architecture and Biological Reactions", in: Biological Science, 24 (10), pp. 1375-1377, here p. 1376; Pauling (1948): The nature of the forces between large molecules of biological interest, in: Nature, 161 (4097), pp. 707-709.
62   Kay (1993): A Molecular Vision, p. 165.
63   Pauling/Delbrück (1940): Intermolecular forces.

the enzymatic synthesis of molecules."[64] Until the mid-1940s, Pauling's contributions to the subject of stereochemical complementarity as the prototypical cause for specificity and its importance for biology increased and he promoted the idea that the study of complementarity was the solution to important biological and medical problems in general.[65]

As previously sketched, the idea that the stereochemical fitting of molecules and their biological functions in organisms were strongly related had already been mentioned by Emil Fischer and Paul Ehrlich in the late 19th century. Pauling's novel contribution, however, was the emphasis on the importance of molecular complementarity for *all* biological phenomena in the framework of what he considered.[66] The scope of his projects thus went beyond attempts to come up with a theory of protein folding that also accounted for antibody determination. Rather, Pauling aimed at formulating "a general theory of the molecular basis of biological specificity."[67] In "Molecular Architecture and Biological Reactions" (1946), Pauling refers once more to the lock-and-key analogy in order to explain his views on specificity and its physico-chemical basis, molecular complementarity:

"This explanation of specificity in terms of lock-and-key complementariness is due to Paul Ehrlich, who expressed it often, in words such as 'only such substances can be anchored at a particular part of the organism which fit into the molecule of the recipient combination as a piece of mosaic fits into a certain pattern.' In recent years the concept of complementariness of surface structure of antigen and antibody was emphasized by Breinl and Haurowitz, Mudd and Alexander, and then was strongly supported by me in the course of an effort to understand and interpret serological phenomena in terms of molecular structure and molecular interactions. Since 1940 my collaborators ... and I have gathered a great amount of experimental evidence about antigen-antibody interaction, which not only supports the general thesis that

---

64 Pauling/Delbrück (1940): Intermolecular forces operative in biological processes, p. 78. See also Kay (1993): A Molecular Vision, p. 173.
65 Kay (1993): Molecular Vision, p. 164-168.
66 Kay (2000): Who wrote the book of life, p. 50.
67 Pauling (1975): The molecular basis of biological specificity, in: Nature, 248, pp. 769-771, p. 769.

serological specificity is the consequence of structural complementariness, but provides information about the extent of complementariness."[68]

Interestingly, Pauling refers once more to Paul Ehrlich and not to Emil Fischer as the founder of the lock-and-key analogy. Moreover, the hypothesis of molecular complementarity in terms of lock-and-key complementarity as the basis for serological specificity has at this point been turned into an experimental research program, conducted by Pauling and his coworkers at the California Institute of Technology. The emphasis on molecular size and shape as the determining factor for specificity rather than merely on the chemical properties of molecules became a key element of this program and of Pauling's overall agenda.[69] Using this framework raised and legitimized an important research problem for chemical research on macromolecules involved in specific molecular processes that form the basis of biological processes, namely the necessity to uncover the spatial structure of the respective complementary interactants.[70] According to Pauling, there was strong

---

68 Pauling (1946): Molecular Architecture and Biological Reactions, p. 1376.
69 See e.g. Pauling (1945): Molecular structure and intermolecular forces, in: Landsteiner (ed.): The Specificity of Serological Reactions, Cambridge (Ma), pp. 275-293; (1946): Molecular Architecture and Biological Reactions, in: Biological Science, Volume 24 (10), May 25, p. 1376; (1946): Molecular Architecture and Medical Progress, Radio talk broadcast on the New York Philharmonic-Symphony Radio Program sponsored by the U. S. Rubber Co., October 13, New York; (1946): Molecular architecture and biological reactions, in: Chem. Eng. News, 24, pp. 1375-1377; (1948): The nature and forces between large molecules of biological interest, Friday evening discourse at the Royal Institution of Great Britain, London, on February 27, in: Nature (161), pp. 707-709; (1948): Molecular Architecture and the Processes of Life, 21st Sir Jesse Boot Foundation Lecture, May 28, 1948, Nottingham, England, pp. 1-13; (1948): Molecular structure and biological specificity, Presidential address at Section 2, 11th International Congress of Pure and Applied Chemistry, London, July 17-24 (1947), in: Chem. Ind. Supple, pp. 1-4.
70 Pauling (1946): Molecular Architecture and Biological Reactions, in: Biological Science, Volume 24 (10), May 25, p. 1376 and Id. (1948): The nature and forces between large molecules of biological interest, Friday evening discourse at the

"evidence that the specificity of the physiological activity of substances is determined by the size and shape of molecules rather than primarily by their chemical properties, and that the size and shape find expression by determining the extent to which certain surface regions of the two molecules are complementary in structure."[71]

The hypothesis that molecular shape was the distinctive feature of biological specificity reactions legitimized X-ray diffraction methods for an audience that was interested in physiological and medical issues. For instance, in his 1949 lecture "Structural Chemistry in Relation to Biology and Medicine" at the City College Chemistry Alumni Association, Pauling argues that further development in the study of "detailed molecular structure" by means of X-ray-analysis will enhance medical diagnosis and care significantly:[72]

"It is my belief that we are just on the verge of taking this great step forward, that during the next ten or twenty years very significant progress will be made in the development of a real science of pharmacology, and that there will be great progress in

---

Royal Institution of Great Britain, London, on February 27, in: Nature (161), pp. 707-709, here p. 708.
71 Pauling (1946): Molecular Architecture and Biological Reactions, in: Biological Science, Volume 24 (10), May 25, p. 1376.
72 Pauling (1949): Structural Chemistry in Relation to Biology and Medicine, in: Baskerville Chemical Journal, 1 (1), pp. 4-7. This is just one of many similar sources in which Pauling envisions the great future of medicine that could be achieved by the extensive study of molecular interactants, in particular by deepening the knowledge of molecular shape. For other talks and articles that address the same issues, see e.g. Pauling (1946): Molecular Architecture and Medical Progress; (1946): Molecular architecture and biological reactions; (1948): Molecular Architecture and the Processes of Life; (1948): Chemical achievement and hope for the future, Silliman Lecture presented at Yale University in October, 1947, on the occasion of the Centennial of the Sheffield Scientific School, in: American Scientist, 36, pp. 51-58, and Pauling/Itano/Singer/Wells (1949): Sickle cell anemia, a molecular disease, in: Science, 110, pp. 543-548. For Pauling's influence on 20th century medicine, see Strasser (2006): A World in One Dimension.

medical practice. In order to achieve this end we must learn three things: the detailed molecular structure of the chemotherapeutic substances, the detailed molecular structure of at least some molecular constituents of the organisms against which they are directed ...and of the human organism, with which the agents come into contact, and the nature of the forces involved in the intermolecular interactions between the substances and the organism. [....] Since 1913 the X-ray diffraction method and other methods of investigation of the structure of molecules made it possible to find out exactly how atoms are attached to one another in molecules such as those of the sulfa drugs and penicillin."[73]

To summarize, Pauling used the concept of complementarity to argue for the importance of little changes in the configuration of macromolecules and their effect on biological functions. Although he ascribed the origin of this idea to Paul Ehrlich, it is Fischer's interpretation of lock-and-key complementarity on which Pauling's approach was based. Similar to Fischer, Pauling understood complementarity as a relation between molecular configurations and highlighted the role of small configurational changes. Beyond that, he provided a physical interpretation of the causes of lock-and-key complementarity.

4.4 THE LOCK-AND-KEY ANALOGY IN SCIENCE ADMINISTRATION AND CROSS-DISCIPLINARY COMMUNICATION AT CALTECH

Between 1940 and 1960, Pauling's view of complementary reactions as causes of fundamental biological and medical processes was turned into a large-scale biochemical research program dedicated to "the attack of fundamental problems in biology and medicine."[74] A group of chemists and biologists[75] surrounding Pauling at Caltech realized several projects on the physical and chemical study of biological phenomena, all of which were

---

73 Pauling (1949): Structural Chemistry in Relation to Biology and Medicine, p. 4f.
74 Kay (1993): Molecular Vision, p. 165 and p. 176.
75 Among them were the embryologist Albert Tyler, the geneticists Alfred Sturtevant and Sterling Emerson, and the immunologist Dan H. Campbell. See Kay (1993): Molecular Vision, p. 176-182.

funded by the Rockefeller Foundation (RF). As Robert Kohler and Lily Kay have shown, the institutional setting of these projects carried out jointly by chemists and biologists had a profound impact on the formulation of scientific problems and on the overall scope of research by means of general grant policies.[76] Caltech's close ties to the RF, especially the collaboration between Pauling and Warren Weaver, the director of the Natural Science Division of the RF, played a central role in the institute's their biochemical success story.[77]

4.4.1 The making of a research agenda – Caltech's biochemistry department and the Rockefeller Foundation

At the very beginning of his administration period as director of RF's Natural Science Division, Warren Weaver proposed a funding agenda that included a systematic plan of how to construct a long lasting influence on America's scientific profile. In order to meet this goal, Weaver suggested providing large funds for a few selected fields that "contribute in a basic and important way to mankind."[78] As for the decisive funding criteria, Weaver noted that

"[i]t is proposed for the future program that interest in the fields in question be the dominant role in the selection process. [...] The choice of fields of interest is influenced by several considerations. The field must contribute in a basic and important way of mankind; it must be sufficiently developed to merit support, but so imperfectly developed as to need it; it should be a field in which the contributions of the Foundation will play a critical role in producing and stimulating development that otherwise would not occur within reasonable time."[79]

---

76 Kay (1993): Molecular vision, pp. 8-11; Kohler (1991): Partners in science, Chapter 10.
77 Kay (1993): Molecular Vision, pp. 149-190.
78 The Rockefeller Foundation (December 13, 1933): "The natural and medical sciences cooperative program." (RAC, RG 3.1, Series 915, Box 1, Folder 7).
79 The Rockefeller Foundation (December 13, 1933): "The natural and medical sciences cooperative program."

Pauling's proposed project "on the structure of hemoglobin and related substances" (1934) at Caltech seemed to have met exactly these conditions.[80] Weaver, who was excited about Pauling's outstanding reputation in physical chemistry, encouraged him to stick to problems with medical significance, and did everything he could to generously support his plans studying the fundamental problems in biology and medicine from a chemical perspective.[81] Rumors about Weaver's directional funding policy started to spread in 1937 within the board of the Rockefeller Foundation. One of the trustees, Herbert Spencer Gasser, an established physiologist and director of the Rockefeller Institute for Medical Research between 1935 and 1953, questioned Weaver's objectivity in the distribution of grants, and accused him of following "a plan" by supporting Caltech's organic chemistry section in such an extensive manner.[82] In 1938, another trustee, Roger Adams, complained about giving $300,000 to the California Institute of Technology in order to strengthen their organic chemistry staff.[83] Weaver reacted to Adam's concern with a detailed letter, in which he defended the support of "the bio-organic field" in general and explained that "the California Institute grant represents a frank attempt to initiate the development of bio-organic chemistry in an otherwise very strong institution, the plan involving most promising and favorable connections with the existing strengths in other fields."[84] In the same year, Weaver consulted Pauling several times as an expert in protein chemistry and asked for material that would express the overall significance of Pauling's research at Caltech. In reaction to this in-

---

[80] Letter from Pauling to the RF, November 22, 1934 and letter from Weaver to Pauling, November 23, 1934. (Warren Weaver Papers, RAC, Series 1.2, Box 10, Folder 136).

[81] Letter from Weaver to Pauling, December 19, 1933. (Warren Weaver Papers, RAC, Box 10, Folder 136). See also letter from Weaver to Pauling, October 3, 1934 (Box 10, Folder 136).

[82] Warren Weaver Diaries, entry from November 6, 1937 (RAC). Robert Kohler also mentions Gasser's distaste for the support of "plans" and "programs" (See Kohler (1991): Partners in science, p. 388.).

[83] Letter from Roger Adams to Weaver, April 6, 1938 (RAC, RF, RG 1.1, Series 205, Box 6, Folder 79).

[84] Weaver's response to Roger Adams, April 13, 1938 (RAC, RF, RG 1.1, Series 205, Box 6, Folder 79).

quiry, Pauling sent Weaver two drafts of papers, "The Molecular Structure of Proteins" and "Hemoglobin and Magnetism",[85] with a note saying:

"I do not have any published lectures on non-technical discussions of our protein work at Caltech, but I am enclosing some material which I hope you can use in preparing your statement. I note that the style of my material is not uniform, because part of it is copied from some old notes and part is freshly written, but I hope that this does not trouble you. I shall send later some x-ray photographs which might be suitable for illustration. [...] I would be glad to write more about any of these subjects if you desire it [...]."[86]

This letter indicates that Pauling – by arranging and adapting part of his scientific publications as well as by offering to write popular science articles – was willing to support Weaver's overall agenda at the RF, which, as one might suspect, was also intended to strengthen Pauling's position in front of the RF board of trustees. Apart from complying with the documentation and information demands of his primary funder, it also shows that by supporting Weaver, Pauling enhanced the opportunity for gaining further funding for his research. Furthermore, the letter presented above also indicates that Pauling did not explicitly work on "non-technical discussions" of his protein studies until he supported Weaver in the protein campaign in 1939, or at least that he himself did not consider these earlier papers as "non-technical".

In the early 1950s, Weaver saw once more the necessity of promoting Pauling's research on proteins within the RF board of trustees and directors. After a conversation with Joe Willits, the director of Rockefeller's social science division, Weaver started to work on a pamphlet that explained to a lay audience why Pauling's protein work at Caltech was so important.[87]

---

85 Letter from Pauling to Weaver, August 11, 1939. (RAC, RF, RG 1.1, Series 205, Box 6, Folder 82).
86 Ibid.
87 See Weaver's manuscript of "Why should Joe Willits care about the structure of proteins?", Inter-Office Correspondence, September 20, 1951 (RAC, RF, RG 1,2, Series 205, Box 4, Folder 27).

Weaver's pamphlet argued for the importance of chemical protein research for insights into disease and even life itself. What is crucial is that Weaver made explicit reference to the lock-and-key analogy in connection with the role of molecular geometry for biological processes:

"[T]he specificity of a protein molecule is understandable only through a knowledge of the actual space structure of the molecule. Various ideas have been developed in protein chemistry about molecules 'fitting' in their surface configuration. One hears the language of a 'key and lock' combination. Indeed, even the biological phenomenon of reproduction has been at times described in terms of an ultimate molecular process which involves a protein 'template' or "mould" which is used to reproduce a basic pattern of life."[88]

In line with Pauling, Weaver argued that the physical and chemical analysis of protein molecules would form the basis for addressing biological and medical problems. Furthermore, Weaver's and Pauling's correspondence shows that both worked on a plan to communicate these programmatic ideas to a non-scientific audience, in this case the financiers of the Rockefeller Foundation.

As the above quote exemplifies, the complementary fitting of molecular configuration was explained by recurring to the lock-and-key analogy. Another interesting aspect is the linkage of the lock-and-key analogy and the idea of a template or mold as pattern for reproduction that was considered to be one of the most important puzzles for the life sciences at that time.

The usage of the lock-and-key analogy was, however, not restricted to contexts of campaigning and politically effective popularization. It also served as a framework to establish a link between several chemical and biological research projects under the umbrella of Pauling's large-scale biochemical program on the study of specificity reactions by means of physical and chemical methods in the 1940s and 50s.

---

88 Ibid.

4.4.2 Analogy-based projects and the construction of lock-and-key-like models at Caltech

Immunochemistry was one of Caltech's pillars in the 1940s and 50s.[89] Especially Pauling's and David Pressman's research project on the "Experimental Investigation of the Structure of Antibodies and the Nature of Immunological Reactions" held large promises for the possibility of controlled antibody synthesis, which in turn gained vast importance for medical care during the war.[90] As Kay notes, the RF provided Caltech with $33,000 for the period 1941-44 to develop their program of "immunochemistry in conjunction with serological genetics."[91] As a result, the immunochemical staff and laboratory equipment expanded and immunological methods were applied to other biochemical fields.[92] Immunochemical research was further fostered by some of Caltech's geneticists, e.g. Alfred Sturtevant and Sterling Emerson, who were not primarily working on immunochemical phenomena, but interested in the similarities between genes and antigens.[93] Yet, the resulting cooperation between biologists and chemists in the cutting edge area of immunochemistry was also strongly motivated by the prospect of huge grants from the RF. As Kay notes, Caltech's biology division had some drawbacks during the late 1930s and 1940s and with George W. Beadle's departure to Stanford in 1937, the innovative spirit, for which Caltech's biology staff was known in the early 20th century, was weakened.[94]

The idea that all kinds of biological specificities could be understood in terms of "antibody-antigen-like" chemical interrelations was emphasized several times in Caltech's proposals and reports to the RF and to the Foundation of Infantile Paralysis in the mid- and late 1940s.[95] At the same time,

---

89 Kay (1993): Molecular vision, p. 136 and p. 168-177.
90 "A proposed project of experimental investigation of the structure of antibodies and the nature of immunological reactions", March 18, 1941. (RAC, RF, RG 1.1, 205, Box 7, Folder 92).
91 Kay (1993): Molecular vision, p. 185.
92 Ibid.
93 Ibid., p. 170 and Id. (2000): Who wrote the book of life, p. 52.
94 Kay (1993): Molecular vision, p. 167.
95 See e.g. the "Rockefeller Foundation Proposed Program of Research 1945-46" (Caltech Archives, Biology Division, Box 62, Folder 13); the "Rockefeller

the linkage between the enzyme- substrate and antibody- antigen relationships was preserved; both were used interchangeably and served as a model for chemical complementarity and its causal powers in other fields of biology, e.g. virology, bacteriology, embryology and genetics.[96] This reference by analogy became an integral and successful funding strategy of groups around Linus Pauling, which, in turn, gave other Caltech scientists the chance to collaborate and participate in one of the most generously supported programs in the US.[97] This becomes visible most strikingly by the usage of the enzyme-substrate, respectively antibody-antigen analogy in grant-related proposals and reports of serological genetics and embryology, which were assimilated by Pauling's expanding biochemical program by the mid-1940s.[98] According to Kay, two of the leading figures in embryology and serological genetics at Caltech, Albert Tyler and Sterling Emerson, completely changed directions after their encounter with Pauling and his program on macromolecules.[99] As will be clarified in the following paragraphs, from the mid-1940s on, Albert Tyler's and Sterling Emerson's research was described in terms of enzyme-substrate and antibody-antigen analogies in proposals and reports to the RF. The tendency to depict and explain not only the operative kind of specificity, but almost every research activity in these fields with antibody-antigen terminology increased by 1945 after projects in serological genetics and embryology had become an

---

Foundation Proposed Program of Research; Chemistry and Biology Progress Report 1946-47"; the 1947 report of Caltech's Biology Division "Chemical Biology at the California Institute of Technology", p. 29-31. (Caltech Archives, Biology Division, Box 22, Folder 12); see also the 1949 "The Rockefeller Foundation: Confidential Monthly Report. For the Information of the Trustees", January 1 (Nr. 108), 1949: p.1-16 (Chemistry joins Forces with Biology), here pp. 5ff (Caltech Archives, Biology Division, Box 22, Folder 12).

96 See Kay (2000): Who wrote the book of life, p. 49.
97 See Kay (1993): Molecular Vision, p. 185.
98 By 1945 projects on serological genetics and immunochemistry were integrated in the jointly written grant proposals of Caltech's Chemistry and Biology Division. See e.g. the "Rockefeller Foundation Proposed Program of Research; Chemistry and Biology Progress Report 1946-47" (Caltech Archives, Biology Division, Box 62, Folder 13).
99 Kay (1993): Molecular Vision, p. 185.

explicit part and a key element of Pauling's large-scale biochemical research program.

Beginning in 1938, the RF had supported Tyler's and Emerson's work on a small-scale basis, but never indicated long-term support.[100] The comparison of embryological, genetic and immunological specificities, as well as the conviction that immunological and serological methods would enhance the work in embryology and genetics were already part of these earlier proposals and reports to the RF. In the 1940 report about RF-funded research in genetics and, in part, embryology,[101] the application of immunological theories and techniques to genetics was introduced as "the best hope of attacking the general problem of gene action."[102] The conviction that

---

100 See Tyler's report "Of the work supported in part by a grant from the Rockefeller Foundation during the year starting July 1, 1939" (RAC, RF, RG 1.1, series 205, Box 6, Folder 83). In the acceptance letter of the RF grant for the project on "Biological Applications of The Principles of Serology" in 1940, A.H. Sturtevant (Chairman of Biology Council at that time) assures the RF that the "responsibility for maintenance of the entire project at the end of the three-year period" is no more than would be inevitable, and should mean no increase in the staff, and after the installation of new equipment on this grant the work of present members of the staff should be no more expensive than it would otherwise be." (Letter from A.H. Sturtevant to Barrett, June 27, 1940; Caltech Archives, Biology Division, Box 4, Folder 10).

101 "Biological Applications Of The Principles of Serology", Research Report (1940). Caltech Archives, Biology Division, Box 4, Folder 10.

102 "One of the promising fields of study is the application of the techniques of immunology to genetics. The results already obtained here have led Haldane and Irwin to suggest that antigens may often be rather direct gene products. It seems likely that in this field lies the best hope of attacking the general problem of gene action. In any case the kind of specificity which the geneticists and immunologist both deal with is so similar that use of both techniques on the same material is certain to yield valuable result. There are a whole series of general biological problems that are in need of study by immunological methods; some of these are already under investigation at the Institute, and we hope that others may be undertaken if adequate equipment is available." ("Biological Applications Of The Principles of Serology", Research Report (1940), p. 1 [Caltech Archives, Biology Division, Box 4, Folder 10]).

such a technological and intellectual transfer could be successful was legitimized by the similarity of specificity reactions: "[I]n any case, the kind of specificity which the geneticists and immunologist both deal with is so similar that use of both techniques on the same material is certain to yield valuable result."[103] As part of this grant report, the description of Albert Tyler's project on the "fertilization reaction in marine eggs" was explicitly oriented towards the phenomenon of specificity and its occurrence in fermentation and serological, respectively immunological, processes. In these earlier proposals and reports, Tyler's approach is described as an extension of the work of Frank Rattray Lillie, a well-established American zoologist and embryologist who advocated the relation between serology and embryology in the first decade of the 20th century. According to the 1940 RF report, Lillie

"called attention to the parallelism in specificity of serological reactions and in fertilization, and studied agglutination of sperm by egg-water in attempts to push the analogy further. Recently Tyler, in our laboratory, has been studying this and related phenomena, and has obtained promising results (partly in confirmation of earlier studies by other workers)."[104]

Tyler studied the necessary conditions for the agglutination of sperm and egg in different species. Following Lillie, he assumed that fertilization takes place when sperm and egg have a relationship of specificity; the agglutination of sperm in egg-water and the presence of a substance called agglutinin were taken as an indicator for this kind of specificity.

In the 1940 RF report, Tyler emphasized that the agglutination reaction between sperm and egg was in many ways similar to the specificity reactions observed in the realm of immunochemistry.[105] The interpretation of sperm-egg agglutination in terms of (inter-)relationships like those found in immunology, mainly raised three research tasks: The determination of "chemical and physical properties of the specific substances concerned, the

---

103 Ibid.
104 "Biological Applications Of The Principles of Serology", Research Report (1940), p. 1f. (Caltech Archives, Biology Division, Box 4, Folder 10).
105 Ibid, p. 2f.

degree of species specificity, and the nature of certain inhibiting substances found".[106]

One could argue that the application of the concept of immunological specificity to the phenomenon of fertilization drove Tyler's research into a chemical direction as he mainly concentrated on the properties of substances involved in the fertilization process. However, one must take into account that Tyler's research of immunological-like reactions in fertilization processes was conducted in a rather classical immunological manner in that the main questions of interest concerned the degree of species and tissue specificity.[107] Thus chemical analysis was rather a means of dealing with classical immunological problems than an end in itself.

What the preceding analysis of Tyler's project descriptions in proposals and reports shows, is that in the early 1940s his research was already characterized as the study of specificities and that immunology, respectively immunochemistry, was used as a prime example for the successful study of the phenomenon of specificity. However, Tyler was not explicitly interested in the physico-chemical basis of embryological specificity at this point; neither complementarity, nor shape were an issue.[108] Kay interprets these findings such that Tyler used immunochemical concepts mostly as "window dressing" for funding proposals.[109] Yet, she also mentions that his "ideas stimulated Sturtevant to explore the chemistry of mutations in a novel way" and eventually led to the proposal of an hitherto unknown mechanism for gene action.[110]

Clearly, Pauling's reputation at the institute and the impact of the template theory of antibody formation on the field of immunology played a role in Tyler's and Emerson's explicit orientation towards "analogous specificities" and their chemical elucidation. The interest in the chemical study of

---

106 Ibid.
107 See e.g. Tyler's 1940 article on "Sperm agglutination in the keyhole limpet", in: The Biological Bulletin, LXXVIII (2), p. 173ff.
108 This can also be supported by Lily Kay (1993) who mentions that Tyler had worked on "classical problems in embryology since the 1920: fertilization and the development of marine vertebrates." (See Kay [1993]: Molecular Vision, p. 168).
109 Kay (1993): Molecular Vision, p. 167.
110 Ibid.

phenomena of specificity and the relation between fields like immunology, serology and embryology, which seemed to be dealing with similar kinds of specificities, was, however, nourished already by the work of Landsteiner and Lillie in the early 20th century, which was long before Pauling entered the field. Nonetheless, efforts to combine these studies under the umbrella of a program at Caltech that searched for complementary substances were not made until 1942. There is evidence that this cooperation was motivated by the active role of RF officers. In December 1942, Frank Blair Hanson, former RF president and the Natural Science Division's associate director at that time, suggested a closer cooperation between different groups working on problems related to immunology and showing funding potential:

"We are in the somewhat unusual position at present of having two grants for the support of immunology at the California Institute of Technology. It would not be advisable for us to consider a third and independent grant in the same field, in the same institution. Dr. Sturtevant's grant expires next June, as does also the special one year-grant for immunological studies under Prof. Pauling. While I am not able to give any indication, at present, as to whether further support in this field is possible, it is nevertheless true that the officers would be willing to give serious consideration to a well thought through program in immunology involving the various interests in this field at Cal Tech. I would, therefore, suggest that you use this letter as the basis for further discussion with Professors Sturtevant and Pauling."[111]

---

111 Letter from Hanson to Tyler, December 1942 (RAC, RF, RG 1.1, 205, Box 7, Folder 93). Two years later, on November 2nd (1944), Hanson wrote in a letter to Pauling that it "does not seem feasible to us to consider this new (or at least in some senses new) program for protein research entirely apart from the current program which we are now supporting in immunology. Our suggestion is that after full discussion with your own colleagues and with Professor Sturtevant's group regarding the inter-relationship, and perhaps the fusion, of the two projects, you send us the results of these conferences." In advance to this letter, Pauling had applied for another large grant for his work on the structure of proteins, including antibodies. (Box 7, Folder 95).

In 1943, Tyler started to cooperate more closely with Alfred Sturtevant and Sterling Emerson on problems in immunology, embryology and serological genetics.[112] In a recommendation letter to Hanson, George Beadle, who was at Stanford at that time, described Emerson's part of the project as investigating the theory that the relation "between gene and antigen-antibody [...] could be such that by controlling the production of specific antibodies it might be possible to find a way of inducing specific gene mutation."[113] He ended his letter with a general comment about the immuno-genetic program at Caltech that, according to him, used "the same general theory as a basis" and recommended to "get a toe hold [...] whenever there is an opportunity" since "it cannot be predicted where the break through is going to come in attacking a fundamental problem such as that concerning gene-antigen-antibody interrelations."[114] Tyler's research that went along similar lines concerning the relation between fertilization processes and the antibody-antigen reaction was recommended to Hanson by Frank R. Lillie. Lillie wrote that he "felt the subject was very important", when he worked on the topic himself in 1914, but that "it remained neglected for many years until Dr. Tyler has taken it up very seriously and has found in this reaction a subject that lends itself very well to investigation of fundamental problems of immunology."[115]

These sources indicate that Sturtevant's, Emerson's and Tyler's research was perceived as a joint and Caltech-specific program towards the application of the antibody-antigen analogy to problems of genetics and embryology. However, the novelty of this approach is not found entirely on the epistemological level. Analogies between serological, immunological and embryological processes and reactions were quite common in the early 20th century.[116] What made research along these lines promising and innovative during the mid-1940s was the attempt to combine intellectual and

---

112 Kay (1993): Who wrote the book of life, p. 170.
113 Recommendation letter from Beadle to Hanson, May 24, 1943 (RAC, RG 1.1, Series 205, Box 7, Folder 94 and Caltech Archives, Biology Division, Box 4, Folder 11).
114 Ibid.
115 Recommendation letter from Frank R. Lillie to Hanson, May 20, 1943 (RAC, RG 1.1, Series 205, Box 7, Folder 94).
116 Mazumdar (1995): Species, p. 68ff.

material resources and to thereby expose the general importance of the chemical study of biological specificity.

The research reports from the years 1945 and 1946 indicate that Hanson's advice of collaboration was put into action. In 1945, Emerson's and Tyler's studies on experimental embryology and serological genetics had become part of the larger program towards the "Fundamental Problems of Biology and Medicine".[117] What is more, in the 1946 grant proposal to the RF, the work on immunochemistry, serological genetics and embryology is described as an "interlocked program in biology and chemistry":

"The immunochemical work has developed in close relation to researches in serological genetics carried on at Pasadena, until the present, under the direction of Professor A.H. Sturtevant and Drs. Albert Tyler and Sterling Emerson. This part of the interlocked program in biology and chemistry has now been greatly expanded and strengthened by the appointment, as Head of the Division of Biology, of Dr. George W. Beadle, formerly of Stanford and the outstanding expert of the world in the field of chemical genetics."[118]

This integration on the proposal level had lasting effects on the scope and general orientation of Tyler's and in part of Emerson's research. Tyler and his associates ventured "to assume that all cells are composed of layers of complementary substances that can react with one another in the manner of antigens with their homologous antibodies"[119] and Emerson's work in serological genetics subsequently was said to be "based on the postulate that the specific surface configurations of antigens and enzymes arise from a corresponding specific surface on the gene, and that this surface determines

---

117 "A proposed program of research on the fundamental problems of Biology and Medicine" by the Division of Chemical Engineering of Caltech [for submission to RF and the National Foundation of Infantile Paralysis], 5/17/1946 (RAC, RF, RG 1.2, Series, 205, Box 4, Folder 22).

118 Ibid.

119 "Report on work supported in part by a grant from the Rockefeller Foundation during the year starting July 1, 1939" (RAC, RF, RG 1.1, Series 205, Box 6, Folder 83).

'gene specificity'.[120] Emerson argued that drawing analogies between the enzyme-substrate, respectively antibody-antigen reaction, and the latest findings in serological genetics "will open a whole array of genetics and developmental problems to experimental attack from an entirely new direction."[121]

The changes brought about by the integration of serological genetics and embryology into the larger specificity program become most visible in a project description and corresponding list of a 1946 proposal for an "Alleloplastology Project" promoting the establishment of a new "hybrid field" dedicated to the study of complementary substances.[122] In the tradition of Pauling's proposal terminology, the Alleloplastology Project was sketched as a "long-term project of wide scope" promising the "possible discovery of concepts of broad biological significance".[123] The joint project was based on the assumption that "[a]ll biological processes involve reactions in which the components exhibit, in varying degrees, specificity of interaction."[124] The "typical antigen-antibody reactions of immunology" thereby served as a prototype for the study of other biological specificities in the new field of Alleloplastology:

"In sea-urchin eggs a sub-surface substance has been obtained that reacts in antigen-antibody-like manner with the surface material. We have, then, two complementary substances, one on the surface (the gelatinous coat), the other below the surface of the same cell, that are capable of interaction. We may assume, then, that normally they are kept from interacting by means of the membrane separating them. [...] The presence of two such complementary substances implies that there may very well be more. We can, then, conceive of the cell as of being constructed of a mosaic of substances that are pairwise or multi-wise complementary. The existence of such complementary substances implies that they are synthesized in the cell by the operation

---

120 "Rockefeller Foundation Proposed Program of Research; Chemistry and Biology Progress Report 1946-47", p. 35. (Caltech Archives, Biology Division, Series 205, Box 22, Folder 12)
121 Ibid.
122 Serological Genetics and Embryology (Alleloplastology) Project Outline, p.1. (Caltech Archives, Biology Division, Box 62, Folder 13).
123 Ibid., p. 2.
124 Ibid., p. 1.

of the same kind of mechanism assumed to be operative in the formation of immune antibodies; namely, the determination of the specific structural configuration of large molecules by means of template action at the site of the synthesis. Emerson (1945) has proposed hypotheses illustrating how genetic action in the production of specific antigens, enzymes, etc. may involve this type of mechanism. For the examination of the validity of these concepts it will at first be necessary to investigate cells of various kinds for the presence of complementary substances. Preliminary tests indicate their presence in mammalian blood cells."[125]

The term "Alleloplastology" was suggested by Emerson and referred to the fact that "the common feature" of the phenomena investigated "is that the underlying reactions are characterized by a fitting together of the components by complementary surface configurations, as exemplified in the antigen-antibody reaction".[126] The phenomena by which the field of Alleloplastology is characterized are described as chemical reactions underlying biological processes in which the fit of the components is further explained in terms of the "complementarity of surface configurations".[127] The terminology used in the 1945/46 proposal strongly points to the analogy between antibody-antigen and fertilizin-antifertilizin reactions. Furthermore, what is emphasized is that both kinds of reactions are mainly characterized as interactions between "complementary surface configurations" of chemical substances.[128]

Especially Albert Tyler directed his research towards the analogy of antibody-antigen and fertilizin-antifertilizin reactions in the years after the proposal. In his 1948 article on "Fertilization and Immunity",[129] Tyler did not only depict his research as oriented towards "analogous specificities" to antibody-antigen specificity, but went even further by adopting the interpretation of "complementary surface interactions" as a cause for embryological

---

125 Serological Genetics and Embryology (Alleloplastology) Project Outline, p.16f. (Caltech Archives, Biology Division, Box 62, Folder 13).
126 Ibid., p. 1.
127 Ibid., p.1f.
128 Ibid.
129 Tyler (1948): Fertilization and Immunity, in: Physiological Review, 28 (2), pp. 180-219.

specificity.¹³⁰ As in the 1945/46 Alleloplastology project description, the application of immunological thinking to the field of fertilization was accompanied by the creation or extension of a general terminology for the chemical study of specificity and other related "general biological problems".¹³¹ Establishing an appropriate and influential terminology seemed to be one of Tyler's major goals according to his 1948 article. In this article Tyler wrote about his motivation to "encourage consideration" of the terminology related to the Alleloplastology project.¹³² Until this terminology was generally accepted in the scientific community, according to Tyler, "substances like fertilizin and antifertilizin may be designated mutually complementary substances whose interaction occurs in the manner of antigen and antibody."¹³³ In the same article, however, Tyler states that one should not take the analogy between fertilization and immunity too serious. Speaking of the fertilizin-antifertilizin reaction in terms of an antibody-antigen-like reaction could be misleading, as part of the definition of an antibody is bound to the existence or introduction of a foreign substance, the antigen. In the case of fertilization, such a *foreign substance* is missing.¹³⁴ Tyler therefore suggests to speak of "mutually complementary" substances or to stick to the terminology of "alleloplasts" for the respective reactants and "alleloplastology" proposed by Sterling Emerson.¹³⁵ This exemplifies that Tyler was aware of the limitations of the analogy between the antibody-antigen reaction and those chemical reactions involved in fertilization. He pointed out that both reactions were only similar with respect to certain aspects, like the chemical basis between the fertilizin-antifertilizin and the antibody-antigen reaction.¹³⁶

---

130   Tyler (1948): Fertilization and Immunity, p. 183 and p. 203.
131   Tyler (1948): Fertilization and Immunity, p. 203. See also "Serological Genetics and Embryology" (Alleloplastology) Project Outline, p.16f. (Caltech Archives, Biology Division, Box 62, Folder 13).
132   Tyler (1948): Fertilization and Immunity, p. 203.
133   Ibid.
134   Ibid.
135   Ibid.
136   However, Tyler doubts that complementary substances "occur simply in pairs", as the lock-and-key analogy suggests. Rather, he adopts the hypothesis proposed e.g. by Haurowitz and Heidelberger that the process of antibody determination is to be seen as a dynamic process, or in Tyler's words, that "com-

One can conclude that in Tyler's usage, the analogy between immunological phenomena and those occurring in fertilization had constraints and this is what made it useful; it pointed to a clear research direction, namely the search for and the analysis of complementary substances. This is especially striking, if one considers that in earlier proposals in the beginning of the 1940s, the analogy between fertilization and immunology remained rather vague and suggested a range of similarities between both fields, including the application of methods from immunology to embryology, similar research questions and understandings of the phenomenon of specificity.[137] Thus, while the analogy between fertilizin-antifertilizin and antibody-antigen reactions was at first used broadly, it became more specific in the mid-1940s. At that point, Tyler concentrated on one particular aspect, namely the existence of complementary substances.

To conclude, at first sight the *Specificity Program* at Caltech seemed to exist mainly on the proposal level, that is in research project proposals, reports and letters to funding institutions jointly written by researchers from the Division of Chemistry and Chemical Engineering and the Division of Biology. It is in these documents, however, that Caltech researchers got the chance to emphasize the interdisciplinary character of biochemical projects conducted by a group of scientists with diverse training, theoretical backgrounds and specialties who despite many differences were working on similar overall questions. One may assume that these kinds of expressed commitments to common research goals are likely to be, at least in part, motivated by financial interests. The Rockefeller Foundation, itself shaped by and propagating an ideal picture of basic science as a cooperative and socially relevant enterprise,[138] became a major source of "special funds" for projects related to immunochemical and protein research.[139] In the 1940s, research on different kinds of biological specificities developed into a large-scale program, entertained and preserved through funding proposals

---

plementary substances are capable of combination". He calls this the "auto-antibody concept." (Ibid., p. 204).

137   See "Biological Applications of The Principles of Serology", Research Report (1940), p. 1f. (Caltech Archives, Biology Division, Box 4, Folder 10). See also this study, the present chapter, pp. 142-146

138   Kay (1993): Molecular vision, p. 10.

139   Ibid., p. 176.

and reports as well as through inner-institutional public relation campaigns and journals.[140] In the case of embryology, Kay goes as far as to state that "the newly funded physicochemical-embryology at Caltech combined the desire for grants with the lure of scientific fashion; in its new chemical garb embryology fit well within the molecular biology program."[141]

As previously shown, the *Specificity Program* did not remain on the rhetorical level. In the last section, I have examined the way in which it was conceptualized and implemented within Albert Tyler's and Sterling Emerson's projects on embryology and serological genetics in the 1940s and 50s. The analysis of these projects demonstrated how the wide-range *Specificity Program*, initially created to meet the goals and expectations of funding institutions and scientific authorities, affected the scientific development of smaller groups and individuals at Caltech and vice versa. It has been shown that the lock-and-key analogy was at first used as a political instrument in

---

140 Apart from the Rockefeller's Trustees Confidential reports, Caltech's "News Bureau" regularly published short reports that pointed to the development of new and ongoing research projects and successes (i.e. prizes and accepted grants). The joint projects of the divisions of biology and chemistry were of particular interest in the reports of both divisions and helped presenting the research conducted at the Caltech laboratories as a collaborative and exciting venture. For instance in the report of 1948, the biochemical work is presented as follows: "At the California Institute of Technology, scientists from three separate branches – biology, chemistry and physics – are joined in this new field; and so close is their union that some think it should have a name on its own, perhaps 'Physico-Chemical Biology'. What the scientists are studying is the ultimate particle of life itself – its chemical composition and the rules for its behavior. They are, in effect, trying to do for living matter what the atomic researchers are trying to do for electronic matter." In an announcement from the "Office of Public Relations" in April 29 of the same year, the joint projects in biochemistry are described as "a long range research program [...] aimed at investigating basic problems in biology and chemistry" and concentrating on "unsolved problems in what might be called the field of 'molecular biology." (RAC, RF, RG 1.2, Series 205, Box 4, Folder 25). See also the "President's Report July 1, 1947-1950" (Caltech Archives, Box 60, Folder 3) and the "President's Report 1951-52" (Caltech Archives, Box 60, Folder 5).

141 Kay (1993): Molecular vision, p. 170.

order to mobilize resources (e.g. funding) for research projects. This analogy usage was thus, first and foremost, mainly rhetorical. But at a certain point – I have located that point in the mid-1940s - a way of thinking was implemented in other contexts of scientific practice; the joint activity of writing one proposal after another had an impact on the individual research agendas, even if not all of Pauling's collaborators implemented the analogy in their further reasoning and methodological practice. Ankeny and Leonelli (2016) use the term "research repertoire" in order to grasp the "performative, social, financial, and organizational components involved in the establishment, evolution, and reproduction of particular ways of doing research."[142] The concept suits the present case, as it accounts for diverse contexts in which scientific research takes place and encompasses activities performed in order to organize research groups. For a research repertoire to be successful, "colleagues, peers, and large-scale [...] funders" have to be convinced. One could thus argue that the lock-and-key analogy became a crucial instrument in the development and expansion of the biochemistry group's repertoire.[143]

4.5 POSTPONING A PARADIGM SHIFT AT CALTECH?

In the early 1950s, more and more biochemical groups at Caltech and other influential US institutions like MIT and Stanford began to doubt that the extensive study of proteins was the right way to solve the "riddle of life" and intensified their work on nucleic acids.[144] Responsible for this shift in attention was the discovery that Deoxyribonucleic Acid (DNA) carried genetic material and Watson and Crick's discovery of the double helix in 1953.[145] The assumption that DNA was responsible for protein shape and folding conflicted with the view that the three-dimensional structure of pro-

---

142 Ankeny/Leonelli (2016): Repertoires: a post-Kuhnian perspective on scientific change and collaborative research, in: Studies in History and Philosophy of Science Part A, 60, pp. 18-28.
143 Ibid., p. 21.
144 Kay (1993): Molecular Vision, p. 269.
145 Morange (1998): A History, p. 106ff.

teins was the determining factor for gene specificity and other kinds of specificity reactions in the organism; a view that was successfully developed and promoted by Caltech and the Rockefeller Foundation for at least 15 years until that point.[146] According to Kay, "Beadle and Pauling cautiously accepted the significance of the new developments."[147] After Watson and Crick's breakthrough, they concluded in 1953 that "nucleic acids play as important a part in biology as proteins, especially in [the] study of the structure of the gene."[148] Molecular control, which was assumed to be achievable mainly by the study of proteins until the late 1930s, soon became DNA control.[149] The growing emphasis on genes in all kinds of biochemical research fields can be illustrated by a published research report of Caltech's biochemical groups from 1955:

"Investigations of gene nature and gene function continue to be carried out on a variety of organisms: viruses, bread mold, corn, fruit fly and several other plants and animals. A problem that is currently receiving much attention by geneticists in several laboratories is the so-called fine structure of the gene. It seems probable that in all organisms primary genetic information is carried in nucleic acid molecules in the form of sequences of four subunits called nucleotides. A gene may be defined as a segment of such a molecule that carries the information necessary for defining the total specificity of a protein molecule."[150]

Kay speaks of a clear paradigmatic shift, but also emphasizes certain continuities between the protein and DNA paradigm like the eugenic goal of social and behavioral control and the belief that this could be achieved by the selective support of fundamental research.[151] As Kay points out, Pauling's group did not have any problem in modifying the claim of the fundamental importance of the molecular study of proteins into claims that supported the molecular analysis of nucleic acids. However, what the conclusion of a par-

---

146 Kay (2000): Who wrote the book of life, p. 56f.
147 Kay (1993): Molecular Vision, p. 271.
148 Ibid.
149 Ibid., p. 276.
150 "A short blub for the president" (Research report of the year 1955), Caltech Archives, Biology Division, Box 60, Folder 2, Section on "The gene", p. 8.
151 Kay (1993): Molecular Vision, p. 277.

adigmatic shift fails to acknowledge is, first, that Beadle's team and a large share of the biology staff at Caltech had long worked on other problems than protein research; many geneticists at Caltech followed, beyond the proposal level, a different approach than the protein group.[152] Thus, it remains an open question how sudden the change at Caltech really was, at least from the biologists' point of view. Secondly, the promotion of the significance of protein research at Caltech did not suddenly come to an end in the 1950s. It has previously been shown that Caltech's biology and chemistry divisions had worked on an integration of both fields since the late 1930s and that the studies on proteins served as a linkage point between both fields, as demonstrated earlier in this chapter. This did not suddenly change after the discovery that DNA was the "fundamental pattern of life".[153] Caltech kept its focus on proteins and promoted it continuously.[154] Yet, one of the rather obvious signs for a paradigm shift is the establishment of a new vocabulary of genetic information in the early 1950s. However, as Kay herself mentions, the concept of specificity and the related "discourse of organization" were neither fully, nor immediately replaced by the "discourse of information":[155]

"While it is true that 'information' was often interchanged with 'specificity', the complex displacement was only partial. For studies of static structures – crystallography, three dimensional molecular folding, or material compositions of proteins and nucleic acids – informational representations seem to have been of little conceptual relevance and experimental appeal."[156]

---

152 Examples of these different approaches are e.g. Delbrück's phage group starting in 1947, and Beadles and Emerson's research on genetic recombination in Neurospora. (See Holmes [2004]: Patterns and stages in the careers of experimental scientists, New Haven, p. 38 and Morange [1998]: A History, p. 25ff.).
153 Kay (1993): Molecular Vision, p. 277.
154 One of the fields in which the "protein view of life" (Kay [1993], p. 104) was preserved was medical chemistry and especially Pauling's work on "molecular diseases", such as sickle cell anemia. See Strasser/Fantini (1998): Molecular Diseases and Diseased Molecules, in: History and Philosophy of the Life Sciences, 20 (2), pp. 189-214.
155 Kay (1993): Molecular Vision, p. 53.
156 Kay (1993): Molecular Vision, p. 53.

4. Lock-and-key foundations for molecular biology | 171

To conclude, it is left open to what extent the "discourse of information" replaced lock-and-key terminology and thinking in branches of molecular biology after the late 1950s. This question is beyond the scope of the analysis presented here. Until that point, however, the lock-and-key analogy was used as a popular exemplar of what the physico-chemical analysis of macromolecules achieved for mankind and social welfare and gained importance for specialists in different fields of biochemistry as well as for the public understanding of these branches.[157]

---

[157] See e.g. Weaver's pamphlet on the importance of protein research in the attempt to convince the trustees of the Rockefeller Foundation that the ongoing support of Pauling's biochemistry group was reasonable. (Weaver: "Why should Joe Willits care about the structure of proteins?" Inter-Office Correspondence, September 20, 1951, RAC, RF, RG 1,2, Series 205, Box 4, Folder 27). See also this study, the present chapter, pp. 142ff.

5 Lock-and-key-based modeling and its influence on the development of biochemical research programs

Building on the preceding, historically oriented chapters, I will now adopt a more analytical perspective in order to address the questions of what the lock-and-key analogy was used for and how its influence on contemporary biochemistry can be characterized and explained. While this attempt to draw more generalizing conclusions on the influencing role of the lock-and-key analogy on biochemical thought and practice implies, to a certain degree, a departure from the historic specificity of each case, the argument I will present remains bound to the examined material and time period. Nevertheless I am confident that by going one step beyond a descriptive reconstruction of the historical facts in their respective contexts, the findings of my study may provide some fertile insights for further research on the dynamic interplay of analogy usage, modeling, and the making of research programs.

I will argue that lock-and-key analogy usage had an influence on both modeling and on the development of a particular long-term research program, which I will further characterize in the course of this chapter. More specifically, I will claim that via the influence of the lock-and-key analogy on modeling processes, a common link between single, individual research programs was established. The result was the creation and preservation of what I call a *long-term analogy-based research program*.

Section *5.1* will deal with the general characterization of the different roles of the lock-and-key analogy. As has become clear in the previous case studies, I distinguish between the heuristic and the reconstructive roles of the analogy in the considered research programs. By that distinction, I am mainly referring to different goals and contexts of lock-and-key analogy usage. I ascribe a *heuristic* role to the analogy in those cases where it was used as a tool for a certain kind of explanatory problem-solving that involved the application of so-called "heuristic strategies", which will be further explained in *5.1.1*. I call the role of the lock-and-key analogy *reconstructive* in those cases where the analogy was used to re-interpret, modify or expand existing concepts, narratives and models, and if that re-interpretation, modification or expansion had lasting effects on the research practice of individuals or groups, e.g. on theory and concept formation, experimental design, and on modeling processes (*5.1.2*). It is important to note, however, that my perspective on the lasting effects of reconstruction is a diachronic and retrospective one, in that reconstruction processes do not need to influence research practice immediately; in my understanding, reconstruction also influences research practice if the re-interpretation of narratives and models has an effect on the next generation of scientists. Furthermore, a re-interpretation, modification or expansion of concepts, narratives or models can be called a re-construction if it has an influence on the work of individuals or scientific groups from another scientific branch or discipline than the one in which the respective concepts, narratives or models were initially introduced.

Section *5.2* will address the influence of the lock-and-key analogy on modeling in the considered time period. The philosophical literature that has hitherto dealt with the influence of analogies on modeling has primarily focused on the heuristic and constructive roles of analogies for model building processes. In this context, analogies are often seen as a first starting point from which scientists can arrive at further developed and more complex models.[1] As mentioned in chapter 1, this view goes back to Max Black's classic considerations on the interrelations between models, metaphors and analogies.[2] Black's process-oriented perspective on modeling and the idea that a wide range of models in science is actually *analogy-*

---

[1] See e.g. Morgan (2012): The World in the Model, p. 172f; Bailer-Jones (2000): Scientific Models as Metaphors, p. 196.

[2] Black (1962): Models and Metaphors, p. 239.

based has attracted many philosophers in the last decades.[3] Terms like "analogy-based modeling", "analogy-based science" or "analogy-based reasoning" have become important phrases in debates concerning the ubiquity and importance of analogies in epistemic processes. Especially the cognitive dimension of analogy usage, that is their role for scientific reasoning, has been emphasized in this respect.[4] One of the most prominent accounts of *analogy-based reasoning* was introduced by Nancy Nersessian who has turned the idea of analogies as tools for a certain kind of thinking into an empirical program.[5] At the core of this program is the attempt to specify the cognitive strategies involved in analogy usage and to distinguish them from other modes of reasoning.

Another philosophical context in which the role of analogies has been emphasized, is the contemporary debate on the general features and functions of scientific models. Here, analogies are mainly treated as the material

---

3 See e.g. Bailer-Jones (2008): Models, Metaphors, and Analogies; Del Re (2000): Models and analogies in science, in: HYLE, 6 (1), pp. 5-15; Frigg/Hartmann (2012): Models in Science, in: The Stanford Encyclopedia of Philosophy: http://plato.stanford.edu/archives/fall2012/entries/models-science/, 01/10/2018, 14:00; Hofstadter (1995): Fluid Concepts and Creative Analogies, New York; Magnani/Nersessian/Thagard (eds.): Model-based reasoning in scientific discovery.

4 See e.g. Abrantes (1999): Analogical reasoning and modeling in the sciences, Foundations of Science, 4 (3), pp. 237-240; Gaboa (2008): In defense of analogical reasoning, in: Informal Logic, 28 (3), pp. 229-241; Alexander/White/Daugherty (1997): Analogical reasoning and early mathematical learning, in: English (ed.): Mathematical reasoning: Analogies, metaphors, and images, pp. 117-147; Knuuttila/Loetgers (2014): Varieties of noise: Analogical reasoning in synthetic biology, in: Studies in History and Philosophy of Science, Part A, 48, pp. 76-88; Nersessian (1988): Reasoning from imagery and analogy in scientific concept formation, in: PSA: Proceedings of the Biennial Meeting of the Philosophy of Science Association, pp. 41-47; Schlimm (2012): A new look at analogical reasoning, in: Metascience, 21 (1), pp. 197-201.

5 See e.g. Magnani/Nersessian/Thagard (1999): Model-based reasoning in scientific discovery; Nersessian et al. (1999): Model-based reasoning in scientific discovery; Nersessian (2002): Model-based reasoning: Science, Technology and Values, and Nersessian (2008): Creating Scientific Concepts.

from which some kinds of models are built or as ingredients, respectively elements, of these models.[6] The dependency relationship between analogies and models has also been used as a demarcation criterion between different kinds of models. Mary Morgan and Daniela Bailer-Jones, among others, use the term "analogical model" for those kinds of models which are in some ways based on or deployed from an analogy. In their characterization of analogical models, both authors stick very closely to Black's view of model building as a linear process that starts with the usage of a "basic analogy" and ends with a systematically deployed model.[7] In contrast to Black, however, Morgan and Bailer-Jones make clear that there are models which are not based on analogies, and that there are modeling processes that work very differently.[8]

To summarize the literature overview, the relationship between analogies and models has often been described such that analogies provide a starting point for model building. As a consequence, analogical modeling, that is modeling by analogy usage, has been widely understood as *analogy-based* model building. I would like to take a step back from this picture of analogy-model interdependencies and shift the emphasis from model construction to *model reconstruction* by means of analogy usage. To be clear: I do not deny that analogies have a heuristic role in model building processes. On the contrary, I would say that this is one of the important aspects that analogies contribute to modeling. However, I claim that analogies have another role in modeling processes, one which has hitherto either been ignored or strongly underestimated in philosophical discussion, namely their *role in model reconstruction processes*. In section *5.2* I will explicate my understanding of model reconstruction and argue that analogies play a dual role in these processes. On the one hand, they provide a link to an established scientific tradition and a well-known scientific context. Using an expression from Christopher Pinckock, I will call this the "anchoring" func-

---

6   Morgan/Morrison (1999): Models as mediating instruments, in: Ead. (eds.): Models as Mediators, Cambridge (Ma), pp. 10-38; Boumans (1999): Built-in justification, in: Morgan/Morrison (eds.): Models as Mediators; and Morgan (2012): The world in the model, pp. 172-204.
7   Black (1977): More about Metaphor, p. 240.
8   Bailer-Jones (2000): Scientific Models as Metaphors, p. 181; Morgan (2012): The world in the model, pp. 79-83.

tion of analogies.⁹ This anchoring function *constrains* the possibilities of analogical model reconstruction to a certain extent in that the resulting model will in some ways be oriented towards an earlier, often canonical scientific context. On the other hand, analogies can be used to *open up* existing models and to adjust them to new findings, ideas and contexts.

Finally, I will call attention to some of the implications of my view on the uses of the lock-and-key analogy for the nature and dynamics of cross-generational research programs. I will propose that the described roles of the analogy designate different stages in the creation, expansion and preservation of a long-term research program. Furthermore, I will argue that the tasks which become important in order to create and preserve such a program are much more about activities of reconstruction and adjustment than about the "open" search for problems and their solutions.

5.1 ROLES OF THE ANALOGY: FROM LOCK-AND-KEY HEURISTICS TO LOCK-AND-KEY RECONSTRUCTION

As we saw in the last chapters, the lock-and-key analogy was used in different ways and for different purposes in the considered individual programs. I have called Fischer's usage of the analogy *heuristic*, first and foremost because it designates that the analogy was used to solve an explanatory research problem in a specific way, that is, by applying certain heuristic strategies.

My understanding of the term "heuristic" was based on Wimsatt's as well as on Bechtel and Richardson's work.[10] In line with these authors, I characterized *heuristic problem-solving* as an approach to the explanation of a certain phenomenon that involves the application of a range of cognitive strategies, such as "simplification", "idealization"/"abstraction", and "incompleteness".[11] What these strategies have in common is that their application leads to a *momentarily* "false" picture of the phenomenon in ques-

---

9   See Pinckock (2012): Mathematical models of biological patterns, p. 482.
10 Bechtel/Richardson (2010 {1993}): Discovering Complexity, p. xxviii-xxxvii, and pp. 23-27, and Wimsatt (2007): Re-engineering philosophy, pp. 76ff.
11 Wimsatt (2007): Re-engineering philosophy, pp. 76ff.

tion. In other words: none of these strategies is about explanatory accurateness. I suggested that Fischer's approach to the explanation of the mechanism of fermentation can be understood as a heuristic problem-solving approach in the sense described above. What is even more important for the present study is my second claim that the lock-and-key analogy had a crucial role in the realization of this approach. This role can be described such that Fischer used the analogy as a *heuristic tool*, that is, as a means to give a simplified, idealized/abstract, and incomplete explanation of the respective phenomenon. Wimsatt, among others, has brought attention to the importance of systematic error production to theory construction. He has argued that heuristic strategies can be used to produce these systematic errors, and thereby can provide a means to control and to learn from them.[12] Hence, due to their (momentary) inaccuracy, heuristic problem-solving strategies can lead to more accurate explanations, because they allow us to make systematic errors and to learn from these.

5.1.1 Heuristic problem solving by means of the lock-and-key analogy

Now let me come back once more to Fischer's heuristic usage of the lock-and-key analogy. I claimed that, in Fischer's program, this analogy served as a heuristic tool in the construction of a model that, in turn, was used to explain a biochemical phenomenon. Specifically, Fischer used the analogy for the explanation of the mechanism of sugar fermentation and for the elucidation of the reaction between enzymes and sugars. He assumed that any chemical reaction between enzymatic and sugar molecules was caused by the complementarity of their spatial structures for which he used the image of a lock-and-key relationship. This guiding assumption was made by analogical inference from his prior experiences in sugar chemistry.[13] Before he turned to the study of fermentation, Fischer had analyzed, classified, and synthesized different groups of sugars and their derivatives. In the course of his studies, the hypothesis that the sugars differed according to the criterion of configuration became more and more likely. By 1891, Fischer was more than confident that the configuration of sugar molecules determined their

---

12 Wimsatt (2007): Re-engineering philosophy, pp. 76ff.
13 See this study, chapter 2, pp. 55-62.

chemical behavior; e.g. their optical activity, melting point etc.[14] The hypothesis that configuration was a decisive factor when it came to the distinctiveness of the single sugars provided a first orientation in his attempt to explain the mechanism of fermentation.[15]

In his first series of experiments on fermentation, Fischer worked with certain types of yeast and observed that the respective sugars which he had used for the experiments were dissolved by the yeast in different ways (e.g. in terms of the reaction time). Based on his prior experience with the sugars, he assumed that their dissolution was influenced by their molecular configuration and even more, that the "yeast cells with their asymmetrically formed agent are capable of attacking only those sugars of which the geometrical form does not differ too widely from that of d-glucose."[16] These results were, however, not really surprising as Pasteur had already suggested in the 1860s that microorganisms would select their food material according to geometrical criteria. Fischer's contribution to this line of research can be characterized as follows: First, he had specified the meaning of "geometrical criteria" by his application of van't Hoff's theory of the asymmetric carbon atom and his usage of the concept of configuration. Furthermore, he had made the concept empirically accessible by introducing formulas that enabled him to classify and compare individual sugars according to their configuration.[17] Thus, in contrast to Pasteur, Fischer referred to geometrical characteristics of the sugars that could be shown and specified in each single case. In short, he had found a way to make the configurational differences visible and this, in turn, provided him with a new semi-empirical instrument for the study of fermentation.[18]

When Fischer started his second series of experiments on sugar fermentation, there was still an open question as to which mechanism the configuration of the sugars influenced the reaction of the yeast. In order to give a

---

14 See this study, chapter 2, pp. 40-52.
15 Ibid., p. 44 and Fischer (1894): Synthesen II, in: Untersuchungen über Kohlenhydrate, p. 108f.
16 Fischer/Thierfelder (1894): Verhalten der verschiedenen Zucker gegen reine Hefen, in: Berichte (27), p. 2037, translated by Lichtenthaler (1994): 100 Years 'Schlüssel-Schloss', p. 2368.
17 See this study, chapter 2, pp. 40-52.
18 Ibid., pp. 47-52.

reliable answer, he changed the experimental setup and used those chemical substances that could be isolated from the yeast, i.e. their enzymes.[19] By using chemical agents instead of microorganisms, Fischer aimed to re-locate the study of fermentation from the biological to the (purely) chemical realm.[20] Based on his first series of experiments, he knew that only some of the tested sugars could be fermented by the yeast cells and interpreted this as the yeast cells being selective with respect to certain geometrical forms of sugar molecules.[21] Furthermore, he suggested that only those molecular forms whose geometry was not "too far" from the geometry of glucose were right for the yeast cells.[22] As for the mechanism of fermentation, Fischer formulated the hypothesis that the enzyme and sugar molecules needed to approximate each other close enough to react chemically. The degree of approximation, in turn, depended on the configuration of both the enzymatic and the sugar molecules.[23] He articulated this hypothesis by means of the lock-and-key analogy: "To make use of an image [...] I shall say that enzyme and glucoside must fit each other like lock and key in order to have any chemical effect on each other."[24] Using the image of a lock-and-key-like relationship to describe the mechanism of fermentation meant to focus on the spatial interaction of two reactants, the enzyme and its substrate. Furthermore, the analogy suggested that enzyme and substrate must be comparable according to the aspect of configuration. Fischer pointed out that what he called the "lock-and-key hypothesis" gave the impulse to expand his project and study not only the influence of molecular configuration on the fermentation of monosaccharides, but to include as well the class of glucosides. He further mentioned that the lock-and-key analogy

---

19 Fischer (1894): Einfluss der Configuration auf die Wirkung der Enzyme, in: Deutsche Chemische Berichte 27, p. 2992.
20 Ibid.
21 Fischer/Thierfelder (1894): Verhalten der verschiedenen Zucker gegen reine Hefen, p. 834, in: Untersuchungen über Kohlenhydrate und Fermente.
22 Ibid., p. 835.
23 Fischer (1894): Einfluss der Configuration auf die Wirkung der Enzyme, p. 2993; Fischer und Thierfelder (1894): Verhalten der verschiedenen Zucker gegen reine Hefe, in: Untersuchungen, p. 829-835.
24 Fischer (1894): Einfluss der Configuration auf die Wirkung der Enzyme, p. 2993; translated by Mazumdar (1995), p. 198.

"provided experimental research with a very particular and tackable problem", namely to study the fermentation of simple asymmetric substances whose constitution and configuration was already well known.[25]

One might ask whether Fischer really needed the lock-and-key analogy to apply the concept of configuration to the realm of enzymology or whether he just used it as a rhetorical device to articulate something he already knew. Or, in other words, what did the lock-and-key analogy contribute to Fischer's approach to fermentation?

First of all, the analogy pointed to the relationship between enzyme and substrate molecules and to the fact that these must fit according to some chemical aspect. At the same time, it left enough room to be filled by a growing stock of hypotheses concerning the influence of configuration on molecular reactions. It is true that Fischer articulated the idea that configuration might play a crucial role in sugar fermentability before he introduced the lock-and-key analogy. Thus, the analogy did not guide him in studying the influence of configuration on fermentability in the first place. However, it added three things to the hypothesis that configuration might be a decisive factor in sugar fermentability: (1) The analogy filled a gap between Fischer's approach to the sugars and his fermentation studies, because the image of a lock-and-key relationship suggested that the enzymes and their sugar substrates must be comparable according to some chemical aspect. This gave Fischer a reason to continue and test his growing stock of sugars from which he knew that they only differed with respect to the criterion of configuration against the respective enzymes. (2) The analogy also suggested that the hypothesis of configurational influence on sugar fermentability could be intuitively justified, as these fitting relationships between two reactants could be found everywhere, even in the most common situations. Finally, (3) using the lock-and-key analogy provided a way for Fischer to link his stereochemical program to the established schools of structural chemistry and to nonetheless distinguish it from traditional approaches to the study of fermentation. Previously used analogies and metaphors (like e.g. die and coin or Pasteur's mirror images) already indicated a mechanical picture of molecular interrelations. The lock-and-key analogy, however, suggested not only that the relationship between enzyme and substrate

---

25 Fischer (1906 {1898}): Bedeutung der Stereochemie für die Physiologie, in: Untersuchungen auf verschiedenen Gebieten, p. 134, my emphasis.

could be understood by the application of mechanical principles; it also pointed into a particular direction of how to access and control the mechanism of fermentation, namely by studying the key (the enzyme) in the same manner as the lock (the respective sugar). This, in turn, meant applying the stereochemical findings in the field of sugar chemistry to enzymology and thus opened the door for the study of enzyme configuration.

To summarize, I am suggesting that the lock-and-key analogy had a heuristic role in Fischer's research program. As I have shown, he used the analogy for explanatory problem solving, specifically, to explain the mechanism of sugar fermentability. By making use of the lock-and-key analogy, Fischer mobilized heuristic strategies, namely simplification (e.g. isolating a particular aspect of fermentation, the configurational relationship between enzymes and sugar), idealization/abstraction (generalizing the relevance of stereochemical configuration from his experience with sugar research for any phenomenon in organic chemistry, and then applying this to the specific phenomenon of fermentation again), and incompleteness (devising an operable model while remaining vague concerning enzyme structure). Thus, Fischer accepted a certain degree of empirical openness in order to make the stereochemical study of fermentation accessible to experimental research.

However, it also needs to be mentioned that the heuristic power of the lock-and-key analogy was limited in two respects: The analogy was neither the only, nor the most important heuristic tool Fischer made use of. His previous experience with sugars and his assumption that the concept of configuration was a decisive factor in fermentability reactions clearly drove his experiments on fermentation in a certain direction *before* he introduced the lock-and-key analogy. Yet, the analogy provided an additional assurance that the study of configurational factors could be a promising approach to the mechanism of fermentation. According to Fischer's own statement, this gave the decisive impulse to test his thoughts on the stereochemical basis of fermentation experimentally.[26] Moreover, it should be mentioned that Fischer used the analogy for a considerably short time peri-

---

26 Fischer (1906 {1898}): Bedeutung der Stereochemie für die Physiologie, in: Untersuchungen auf verschiedenen Gebieten, p. 134. See this study, chapter 2, pp. 58ff.

od and that he did not make use of it in his subsequent study on protein structure.[27]

5.1.2 (Re-)Constructing models and research programs by means of the lock-and-key analogy

Turning now to other programs, one of the striking findings of the case study on Ehrlich's immunological and chemotherapeutic program was that he did not use the lock-and-key analogy as a heuristic tool for explanatory problem solving (*chapter 3*). This was clearly a deviation from the historical literature that hitherto treated Ehrlich's use of the analogy as one of the crucial episodes in the history of the analogy and emphasized its heuristic value in this context.[28] It has been shown that Ehrlich used pictorial representations and a set of theoretical terms that pointed to aspects of the phenomenon in question that were also indicated by the lock-and-key analogy. Just like Fischer, he emphasized the complementary relationship between the molecules of two substances and he also made clear that this kind of complementarity was a major condition for both the possibility and the prevention of bacterial infections.[29] However, molecular complementarity was only one of Ehrlich's research interests in the field of immunology. Another important aspect of his work was the classification of the various substances that were involved in immunological processes according to their supposed functions in these processes. With respect to Ehrlich's heuristic toolbox, there were other, similar pictorial analogies and metaphorical terms that became heuristically powerful in his immunological program, e.g. representations of the "haptophore and toxophore groups".[30] Yet, despite these findings that suggest that the lock-and-key analogy itself did not play a major role in Ehrlich's research, he became known as the scientist who applied the analogy to the realm of immunology and chemotherapy.[31] As I have shown, this can be explained by the powerful role of the retrospective interpretation of his models in different phases of the Ehrlich re-

---

27 Lichtenthaler (1994): 100 Jahre Schlüssel-Schloss, p. 2371.
28 See this study, chapter 3, pp. 72-75.
29 Ehrlich (1900): Croonian Lecture, p. 429.
30 See also the present study, chapter 3.3.2.
31 See the present study, chapter 3.4.

ception between 1908 and the 1940s. In other words, the linkage between Ehrlich and the lock-and-key analogy was, to a great extent, a historical reconstruction by Ehrlich's successors and a result of the reception of Ehrlich's program in Germany and North America. In the phases of reception, the lock-and-key analogy gained vast influence in that the receptor model of immunological and chemotherapeutic reactions was *reinterpreted* in terms of the lock-and-key analogy; it was treated as a successor of Fischer's lock-and-key model of enzyme-substrate relations. I have called this reinterpretation *reconstruction*, as it had an actual impact on how the receptor model was used in the realm of immunology and immunochemistry afterwards, especially in the 1930s and 40s.[32] The crucial point is that the lock-and-key analogy was not used as an epistemic instrument by Ehrlich himself, but rather by others in retrospective communication processes. The purpose of analogy usage in this context was to make sense of Ehrlich's research and to establish a linear biographic path in his scientific achievements. The result was something like a meta-model that subsumed Ehrlich's research programs and the models he created in the course of these programs. I have made clear that this kind of reconstruction in the phase of science reception was performed via communication processes in inner-scientific as well as in public contexts, or, to put it in the words of Ludwik Fleck, this reconstructive activity exceeded the "esoteric circle" of science.[33] Scientists and physicians, as well as science writers and journalists, used the lock-and-key analogy to create a rationale in Ehrlich's scientific achievements. Throughout his scientific career, Ehrlich worked on many different scientific and medical problems, some of them being located in the areas of immunology, bacteriology, toxicology, epidemiology and pharmacology. Despite his various activities in different scientific areas, the Ehrlich reception in the considered time period often focused on models, theories, and hypotheses that could be reformulated in terms of the lock-and-key analogy.[34] In the application of the lock-and-key analogy to different biochemical and biomedical problems, one thus found a common driving force for Ehrlich's way of doing research at the cross-line of biochemistry and medicine.

---

32 I have exposed this in chapter 4.2.
33 Fleck (1979 {1935}): Genesis and development, p. 111.
34 See the present study, chapter 3.4.2.

What is important is that, although Ehrlich did not use the lock-and-key analogy to solve a research problem, it influenced the way in which the next generation of scientists approached research problems in the realm of immunochemistry and the establishing field of molecular biology. This has been shown by my analysis of the lock-and-key analogy in the 1930s and 40s (*chapter* 4). It was in this time period that the analogy became heuristically powerful for antibody research. Marrack, Mudd, and Alexander's framework theory of antibody-formation was built on the foundation of the lock-and-key analogy in that immunological processes were described as "antibody-antigen-like or enzyme-substrate-like" reactions.[35] Pauling, who had begun research in immunochemistry in the late 1930s, modified the framework theory and used it as a basis to formulate a detailed chemical model of how normal globulin chains were modified such that they could perform the function of antibodies in the presence of an immune attack. This model was much more complex than Fischer's lock-and-key model of enzyme-substrate relations with respect to the dynamics of chemical processes involved. However, it still included the lock-and-key analogy and added another dimension to it: The model gave an answer to the question of how "normal" globulin could be turned into a *key* which, in turn, would fit the invading antigen as well as the cells of the host.[36]

In the course of the 1940s and 50s, the lock-and-key analogy was used once more for reconstruction purposes; this time to reinterpret and subsume existing models under Pauling's general view of the molecular basis of biological specificity (*Chapter 4*). Pauling's template model of antibody-formation was expanded and transferred to other related research fields at Caltech; it became the institute's dominant view of molecular reactions underlying all kinds of biological processes (immunity, fermentation, fertilization, cell growth, heredity etc.). A characteristic goal of Pauling's specificity program at Caltech was to *find similarities between smaller projects* that were conducted in the chemical and biological departments of the institution. Finding these similarities was crucial, as it justified long-term support from funding institutions like e.g. the Rockefeller Foundation (RF). The

---

35 This has been shown in chapter 4.2.
36 See e.g. Heidelberger (1939): Quantitative absolute method in the study of antigen-antibody reactions, in: Bacteriological Review, 3 (1), pp. 49-95, here p. 74 and chapter 4.1 and 4.1 of this study.

size, scope and degree of interdisciplinarity of a program was known to be a decisive factor in the distribution of grants from the RF. Here, the lock-and-key analogy provided a tool for the re-orientation of the smaller projects towards similar problems and goals of research and most importantly it led to the unification of these projects. This was exemplified by the proposal for the 'Alleloplastology Project' which tried to integrate several projects with different scientific objects and, in part, different methodological approaches by means of lock-and-key terminology.[37] Furthermore, the analogy was used to explain research problems and strategies to the administrators, the financiers, and the cooperation partners of scientific groups.

In sum: In my interpretation of lock-and-key analogy usage in the first half of the 20th century, I have emphasized two roles of the analogy; its heuristic and its reconstructive role. I have argued that Fischer used the analogy as a heuristic tool for explanatory problem solving. In Ehrlich's program, however, the analogy had a different role; it was used for the reconstruction of Ehrlich's scientific approach and for the retrospective interpretation of his models. In the realm of immunology, the lock-and-key analogy became heuristically powerful (in the sense of an epistemic instrument) in the 1930s where first Marrack, Mudd, and Alexander and then Pauling used it to explain the chemical mechanism of antibody formation. According to my interpretation, the reconstructive phase in which the lock-and-key analogy was linked to Ehrlich's receptor model was a crucial intermediate step before it could gain heuristic power in immunology. After Pauling had used the analogy to articulate his template model of antibody formation in the early 1940s, it was again used in the sense of a reconstructive tool. This time, the reconstruction of models by means of the lock-and-key analogy took place in the context of project management at Caltech. Just as in the reconstruction of Ehrlich's models in the early 20th century, the analogy was now used to subsume different models under the umbrella of a meta-model that would account for a wide range of biochemical phenomena. This meta-model could be called "the lock-and-key model of specificity reactions". In the context of project management, model reconstruction by means of the lock-and-key analogy also had an effect on the level of research practice. Scientists who worked in the field of embryology and genetics – e.g. Albert Tyler, Sterling Emerson and Alfred Sturtevant – tempo-

---

37 See this study, chapter 4.4.2.

rarily adopted the lock-and-key terminology and re-interpreted their findings and models in a way that was consistent with the lock-and-key analogy.[38]

To make my claim on the role of the analogy more intelligible, the table below provides an overview of the considered research programs in which the lock-and-key analogy was used, the phenomena that should be explained in the course of the programs, the entities (substances) involved in these phenomena, and the resulting models. Furthermore, it shows whether the lock-and-key analogy had a heuristic or a reconstructive role in each program and explicates the activities made possible by each role of the analogy and the respective purposes.

---

38 This study, chapter 4.4.2.

Table 1: Overview of the considered research programms in which the lock-and-key analogy was used

Individual research program	Phenomena	Entities	Model(s)	Role of the analogy
Study of fermentation from a stereochemical perspective (Fischer, 1894 – ca. 1907)	Sugar Fermentation	Enzymes & Sugars	The lock-and-key model of enzyme-substrate relations	Heuristic: Realization of a heuristic, explanatory, problem solving approach *Goal: Explanation of the mechanism of sugar fermentation.*
Study of immunological & chemotherapeutic processes (Ehrlich and his reception, 1884-ca. 1930s)	Immunity & Drug action	Antibody & Antigen Receptor & Chemotherapeutic substance	The receptor model of immunological and chemotherapeutic reactions	Reconstructive: Reinterpretation of Ehrlich's scientific work and subsumption of his hypotheses via communication processes in inner-scientific and public contexts. *Goal: Making sense of Ehrlich's scientific work and creating a linear path of his scientific investigations.*

Individual research program	Phenomena	Entities	Model(s)	Role of the analogy
Program on the physico-chemical forces involved in antibody formation and in the linkage between antibodies and antigens. (Pauling, 1938-ca. 1955)	Physico-chemical causes of immunity	Antibody & Antigen	The template model of antibody formation	Heuristic: Modification of the framework theory and creation of the template model of antibody-formation via inner-scientific, trans-disciplinary communication processes. *Goal: Explanation of the physico-chemical basis of antibody formation.*
Caltech's program on the molecular basis of different kinds of biological specificity (The joint bio-chemistry group at Caltech: i.e. Albert Tyler, Sterling Emerson, Linus Pauling, 1945-ca. 1955)	Kinds of biological specificity	Antibody & Antigen Fertilizin & Anti-fertilizin Gen & Antigen	The lock-and-key model of the antibody-antigen-relationship The lock-and-key model of the fertilizin-antifertilizin-relationship The lock-and-key model of the gen-antigen-relation-ship	Reconstructive: Reinterpretation and unification of models from different biochemical branches via communication processes in the project management context. *Goal: Expanding the specificity program, pressumably in order to preserve and rescue (the support of) smaller research projects.*

5.2 ANALOGICAL MODEL RECONSTRUCTION

I have interpreted the historical developments sketched in the previous chapters such, that throughout the first half of the 20th century, a range of similar models have been proposed in different branches of biochemistry, which were all treated as variants of the same basic idea. This basic idea can once more be summarized as follows: The first assumption is that molecules from different substances can complement each other like a lock and a key in their structure, shape, or form. Secondly, it is assumed that the chemical reaction of those "fitting" molecules is responsible for a set of biological functions in the organism, or to put it differently: That the reaction between something analogous to a lock-molecule and something analogous to a key-molecule lies at the heart of fundamental life-processes, as for example fermentation, immunity, sensation, development etc.

The adjustment of different models to the same basic idea, I argue, is a consequence of a certain way of long-term analogy usage. As mentioned above, most philosophers dealing with the relationship between analogies and models emphasize the heuristic role of analogies as starting points for the construction of systematically elaborated models.[39] I have made clear that the heuristic role of analogies is only one of their important functions in modeling processes. Another one, which has not been emphasized in the literature yet and which becomes apparent if one looks more closely at the usage of analogies over a long time period, is their *reconstructive role*.

In the following, I will present some of the consequences for the role of analogies in modeling processes that arise when acknowledging the reconstructive activities that can be performed by analogy usage. In contrast to previously mentioned literature on analogies, I claim that these reconstructive activities strongly influence modeling processes. This is, I argue, because model reconstruction by analogy usage creates similarities between models and can lead to the adjustment of these models towards the same

---

39 See e.g. Bailer-Jones (2000): Scientific Models as Metaphors, p. 196; Magnani, Nersessian and Thagard (1999): Model-based reasoning in scientific discovery; Nersessian et al. (1999): Model-based reasoning in scientific discovery; Morgan (2012): The World in the Model, p. 172f; Nersessian (2002): Model-based reasoning: Science, Technology and Values, and Nersessian (2008): Creating Scientific Concepts; Hofstadter (1995): Fluid Concepts and Creative Analogies.

basic assumptions. In other words: Model reconstruction by means of analogy allows for the subsumption of different models and – given that these models take into account a range of different phenomena – allows for the unification of different areas of research.

The main idea behind the term *analogical model reconstruction* can be grasped as follows: By means of analogy usage, some models can be modified and adjusted to new scientific contexts, problems, ideas, and findings on a regular basis. Understood in this sense, these models can be characterized as *open* tools of scientific investigation. At the same time, model reconstruction by means of an analogy constrains the modeling process in that the product of reconstruction (the resulting model) is interpreted with respect to another context. In the case of lock-and-key analogy usage in the early 20th century, this other context is not only the every-day life experience of the fitting between locks and keys, rather it is a much more specific and scientific context, namely the stereochemical study of sugar fermentation. In fact, I claim that the further we move forward in the development of lock-and-key analogy usage in the 20th century, the more scientific contexts are *suggested and transported* by the analogy. After the analogical model reconstruction process that linked the lock-and-key analogy to Ehrlich's immunological research in the late 19th and early 20th century, the analogy transported not only the context of sugar fermentation, but also a specific immunological context and a certain way of thinking about immunological processes. This transport of scientific contexts was possible, I argue, because the lock-and-key analogy and the respective models influenced each other mutually and enduringly during the process of analogical modeling. The analogy itself was influenced by the scientific context in which it was first used (which was the construction of a stereochemical model of fermentation) and it continued to carry aspects of that context long after. In my interpretation, these aspects were transported and spread via model reconstruction processes. Thereby, the models resulting from these reconstruction processes were influenced by a particular interpretation of the lock-and-key analogy that implied the scientific contexts in which the analogy was previously used. In the following, I will further explicate my view on model reconstruction processes by means of analogy usage.

5.2.1 The open-endedness of model construction processes

My understanding of "model reconstruction" is related to contemporary theories of the processes and strategies involved in model construction and to the idea that model construction is, to a certain degree, an open-ended project. The subject of model construction has been emphasized by, among others, Morgan and Morrison, as well as by Knuuttila et al.[40] The general idea behind emphasizing construction processes is that the way in which models are built tells us something about their special features (e.g. in contrast to theories) and their functions in science. In consequence, the focus on model construction leads to the question of what models are composed of and of how different sources that modelers make use of influence the "nature" of the model or the different techniques of modeling.[41]

As to the question of the composition material, Morgan and Morrison identify theory and data as major sources for the construction of models, while Boumans states that models are composed not only of theories and data, but also of "theoretical ideas, policy views, mathematisations, metaphors and empirical facts."[42] However, both Morgan and Morrison as well as Boumans leave open how exactly the construction of models out of different sources is linked to their scientific roles. Knuuttila and Voutilainen aim to fill this gap by specifying the notion of model construction and by providing an ontology of models as *fabricated, "epistemic artifacts"*.[43]

In line with Morgan and Morrison, Knuuttila and Voutilainen claim that models can fulfill important functions in knowledge generation, because they are constructed in a particular way and that understanding their fabricated nature would allow us to understand their role in science. However, they tell us more about the consequences of a constructivist view of modeling, with respect to the constraints that a specific model building context – and the scientists constructing the model – can pose on the subsequent us-

---

40 Morgan and Morrison (1999): Models as mediators, p. 15; Knuuttilla/Voutilainen (2003): A parser as an epistemic artifact: A material view on models, in: Philosophy of Science, 70 (5), pp. 1484-1495, here p. 1486f.
41 Morgan and Morrison (1999): Models as mediators, p. 15.
42 Boumans (1999): Built-in justification, p. 67.
43 Knuuttila and Voutilainen (2003): A parser, p. 1486.

ages of that model. The core thesis is that models are strongly influenced and somehow "marked" by their construction processes and by the specific goals of the constructor(s). Nonetheless, models are – by merit of their construction – *open-ended instruments*; they can be used for several purposes and in different contexts, they are open to different kinds of manipulation and to a wide range of interpretations.[44]

In what follows, I would like to take Knuuttila and Voutilainen's idea of open model construction one step further and emphasize the diachronic dimension of model construction processes. In contrast to Morgan and Morrison as well as Knuuttila and Voutilainen, I will not only concentrate on the process in which a model is originally built. Rather, I will speak of *model reconstruction* and show how some models are reconstructed on a regular basis by being used and manipulated. What I take from Knuuttila and Voutilainen's account is the assumption that the manufactured nature of models – which I understand as a product of continuous model reconstruction – does not only constrain the models, but also provides them with a certain amount of flexibility. Models are thus flexible in the sense that we can use and re-build them for a wide range of scientific activities and for different scientific problems. Following Knuuttila and Voutilainen, I take some models to be "open-ended" instruments that can be related to new "ideas, instruments and phenomena depending on [their] context of usage."[45] The reinterpretation of Knuuttila and Voutilainen's account with respect to model reconstruction processes allows a new perspective on the dynamics of model building, according to which models are open to scientific change, including change in meaning and with respect to the nature and scope of research problems. In the next section, I will specify a particular type of model reconstruction processes, namely reconstruction by analogy usage, and the changes that occur during these processes. I will deal with the open and constraining roles of the lock-and-key analogy for the construction and reconstruction of biochemical models in the considered programs.

---

44 Knuuttila and Voutilainen (2003): A parser, p. 1485.
45 Ibid.

5.2.2 Reconstruction by analogy, or:
The role of systematic suggestiveness

In this section, I will specify the mutual, bilateral influence exerted by the analogy and the models on each other. My first thesis is that the analogy was linked to the contexts in which it was used for modeling and thus became more and more "loaded" with scientific context in the course of the 20th century. Secondly, I claim that this increasing process of analogy contextualization had an effect on what the analogy suggested and, in consequence, on how it influenced scientific modeling processes from there on. In line with Max Black, I will argue that the influence of an analogy on scientific modeling increases with its "systematic suggestiveness."[46] However, I will make an important extension to Black's usage of this concept. I will propose that the *systematic suggestiveness* of an analogy increases with its historicity and in particular with the number of successful model reconstruction processes for which the analogy was used.

In order to understand what Black means by the term "systematic suggestiveness", let me shortly summarize his view on scientific modeling and the role of analogy in modeling processes: According to Black, modeling is a process that usually starts with the usage of so-called "root metaphors" and "basic analogies".[47] These are often taken from a domain of common sense experience, and end with the formulation of a scientific model by means of analogical inference. Examples for basic analogies would be "light waves" in electrodynamics or the "messenger RNA" in molecular genetics. In both of these cases, the phenomenon in question (light motion or the function of ribonucleic acids in protein synthesis) is explained with reference to common sense vocabulary. The crucial point is that Black makes a distinction between basic analogies and root metaphors on the one hand and elaborated analogies (or models which arise out of analogy usage) on the other. An elaborated analogy is "something more than a simple metaphor."[48] This distinction refers to the kind and degree of suggestiveness, meaning that a basic analogy suggests something in a different way than a

---

46 Black (1962): Models and Metaphors, p. 239. See also Bailer-Jones (2000): Scientific Models as Metaphors, p. 195f.
47 Black (1962): Models and metaphors, p. 239.
48 Ibid.

more systematically elaborated analogy. Basic analogies and root metaphors are "suggestive" in the sense that they put two things into relation with each other (light motion and water motion) and thereby point to those features of the phenomenon in question that can be explained by reference to the other, better known, phenomenon. An elaborated analogy does the same, with the difference that the domain of experience that helps to explain the phenomenon in question provides a complex system of thought; it is "systematically" suggestive.[49] Black does not make this explicit, but taking his considerations into account it seems legitimate to assume that the degree of systematic suggestiveness *increases* with the historicity of an analogy. Or in other words: A basic analogy, which is merely suggestive, *can become a systematically suggestive analogy via a historical process.* In order to explicate my thoughts on this, let me come back to the case study. I am suggesting that in the early history of its scientific usage, the lock-and-key analogy was used in the sense of a basic analogy. When Fischer wrote, "[t]o make use of an image, I shall say that enzyme and glycoside must fit each other like lock and key in order to have any chemical effect on each other",[50] the analogy mainly pointed to a common-sense experience, the fit between a lock and a key; at this point it did not yet "carry" another system of thought, elaborated in another scientific context, and thus may be characterized as a basic analogy.[51] Once the analogy had become systematically

---

49 Black (1962): Models and metaphors, p. 241.
50 Fischer (1894): Einfluss der Configuration, p. 2992; translated by Mazumdar (1995), p. 198.
51 One has to take into account that chemists in the late 19th century were quite experienced with similar metaphors and analogies that were used to describe the mechanical aspects of molecular interrelations (e.g. Pasteur's mirror images or the die and coin metaphor). Hence, one could still argue that, although the lock-and-key analogy itself did not have a scientific history at this point, it referred to a system of thought that was until then elaborated by means of other, similar metaphors and analogies in chemistry. Given that this was the case, we should consider the possibility that the lock-and-key analogy was already systematically suggestive to some extent due to its linkage to other, more elaborated analogies. My point is, however, that the analogy's degree of systematic suggestiveness increased in the course of its uses in biochemical contexts throughout the first half of the 20th century.

suggestive, it did not only point to a common sense relationship (like the one between a lock and a key); it also suggested a system of thought developed within a prior scientific context, namely fermentation chemistry. Both Ehrlich and later Pauling employed the context in which the analogy was first used and interpreted immunological processes in terms of enzyme-substrate processes.[52] It certainly mattered for the suggestive power of the analogy whether this previous context had already gained a high reputation or not. When Pauling used the lock-and-key analogy, both Fischer's chemical model of enzymatic processes and Ehrlich's receptor model had gained canonical status in the developing biochemical sciences. Another implication of the previously sketched view is that systematic suggestiveness also comes with analogy experience; it matters whether scientists know how to use a given analogy. Understood in this sense, an analogy's systematic suggestiveness also refers to pragmatic aspects of scientific practice. For instance, we have to ask, whether scientists are experienced enough in the usage of an analogy or whether they sufficiently trust the analogy in order to deploy it in a systematic way. This was certainly the case in the Caltech episode. In fact, Caltech researchers were even able to use the analogy as an organizing tool in the development of their research projects.[53] One could say that they did not only elaborate systems of thought by analogical inference, they also created working structures by means of analogy usage.

To summarize, I have argued that the systematic suggestiveness of the lock-and-key analogy increased after it has been used in a number of model reconstruction processes. The idea behind this is that every time when an analogy is used to either build or re-adjust a model, it gets loaded with scientific context. Re-adjusting an existing model by means of an analogy can then become a strategy to link one's own individual research program to a canonical scientific context and thereby establish continuity. If we take these cases of model re-adjustment by analogy into account, there are many situations in which scientists can link their models to a well received and sometimes unquestioned scientific context. This, in turn, implies consequences for thinking about the nature and dynamics of long-term research programs. One thing to learn from the present study is that a crucial feature

---

52 This can further be supported by Kay's notion of the protein paradigm, see Kay (1993): Molecular vision, p. 271.
53 See the present study, chapter 4.4.2.

of long-term research programs, namely ascertaining the continuity of core ideas and concepts provided by members of the program,[54] is achieved by a certain way of using analogies. I have proposed that the feature by which analogies can create, expand, and preserve this continuity is their systematic suggestiveness, which, in turn, depends on the contextual content an analogy gains during its use for model construction and reconstruction. Understood in this sense, analogies can become powerful guiding tools for the continuation and expansion of a research program due to their role in modeling processes.

---

54 I have also made this point in Chapter 1.3.

6 Concluding remarks on the construction of analogy-based research programs

In the previous chapters I have analyzed the usage of the lock-and-key analogy in three research programs: Emil Fischer's program on the stereochemical basis of sugar fermentation and enzymatic reactions from 1894 to 1907, Paul Ehrlich's research on immunological & chemotherapeutic processes and its reception from 1884 to the 1930s, and finally Linus Pauling's research on the physico-chemical basis of protein specificity which developed into a large-scale program on all kinds of biochemical specificity reactions, conducted by Caltech researchers in the departments of chemistry and biology from the late 1930s until the 1960s. I have argued that the lock-and-key analogy contributed to the subsumption of these programs under the umbrella of a cross-generational and cross-disciplinary research program on the physico-chemical basis of macromolecular reactions, which can also be seen as one of the corner stones of what Robert Olby has called the "molecular revolution" in biochemistry and genetics.[1] This cross-generational research program can be understood as an analogy-based research program, because an analogy guided the development of the considered individual research programs towards a particular end, namely towards the adjustment of these research programs to other, similar programs and towards the extension of the considered phenomenological scope of the respective programs.

---

1 Olby (1990): The molecular revolution in biology, in: Deby et al. (eds.): Companion to the history of modern science, London, pp. 503-520.

The role of analogies in the making of cross-generational research programs has so far not been a subject of study in the philosophy of science literature. The considerations I have developed based on the three cases examined in this study can therefore provide a starting point for further analysis and theory development. One debate, for which potential linkages are already clearly pronounced, is the discussion on the guiding role of models in scientific development, for which the term "model-based science" has been coined.[2] The general idea is that models are important for research development in many different ways; they fulfill a *guiding role* in scientific processes. Familiar slogans for the guiding role of models are "models as blueprints" (Cartwright), models as "prototypes" (Schaffner), "idealtypes" (Morgan) and "anchors" (Pincock). In all of these accounts the "guiding role" of models stands for a set of features and functions that make models especially useful for the course of a research program or for particularly indispensable scientific activities. The feature of models which is mentioned most often in this respect is their role in providing a system of thought or in putting constraints on descriptions of scientific phenomena by representing specific, limited situations or contexts. Cartwright, for instance, argues that models are constitutive for natural laws. According to this view, natural laws do not exist "out there in the world" without human intervention; instead, situations in which laws hold have to be arranged and designed by models.[3] Another feature that has been mentioned is the flexibility of mod-

---

[2] See e.g. Godfrey-Smith (2006): The strategy of model-based science, in: Biology and Philosophy, 21, pp. 725-740; Gelfert (2011): A model-based representation in scientific practice, in: History and Philosophy of Science Part A, 42 (2), pp. 251-252; Nersessian (2009): How do engineering scientists think? Model-based simulation in biomedical engineering research laboratories, in: Topics in Cognitive Science, 1, pp. 730-757.
The idea that models play a central role in scientific development and that they are particularly related to the dynamics of research programs has already been articulated by classical theories of scientific development in the 1960s and 70s. See Bailer-Jones (2009): Scientific models in philosophy of science, p. 110ff.

[3] Cartwright (1997): Models: The blueprints for laws, in: Philosophy of Science 64 (4), p. S292 -S303, here p. S293.

els regarding the possibility of linking them with other models and thereby using them as patchwork-devices in many different scientific contexts.[4]

Christopher Pincock gives another interpretation of how models, mathematical ones in particular, may be able to guide a research program, namely by providing "anchors".[5] He exemplifies his thoughts with a case study of Hamilton's selfish-herd-model in population biology. Hamilton's model was fundamental for a research program that was later continued and extended by other population biologists. The question of interest for Pincock is why population biologists stuck to this model, although it was insufficiently explanatory in the first place. For one thing, Pincock claims, Hamilton's model provided biologists with a familiar system of thought that enabled them to think more specifically about the phenomenon at issue. This system of thought provided "anchors", assumptions that "link the initial mathematical model to its target systems."[6] Some of these "anchors can be given a realistic interpretation, while others require some sort of idealization."[7] According to Pincock, scientists often stick to their "anchors", because studying other possibilities would eventually force a scientist to abandon those aspects of the research program that proved to be especially fruitful in its beginnings. It is important to note, however, that such "anchors" of a research program are not the mathematical models themselves, but the "assumptions that the scientists in the research program deploy when they specify and investigate their models."[8] These assumptions are based on idealizations in most cases of mathematical modeling (and not on empirical observations), and can be used to unite different models of a research program.[9]

Building on Pincock's idea, I argue that analogies can function as "anchors" during the establishment and expansion of research programs. Just like mathematical idealizations, analogies can be used to unify different

---

4 See e.g. Boumans (1999): Built-In justification, p. 67, and Knuuttila and Voutilainen (2003), p. 1486f.
5 Pincock (2012): Mathematical models of biological patterns: Lessons from Hamilton's Selfish Herd, in: Biological Philosophy, 27, pp. 481-496.
6 Ibid., p. 492.
7 Ibid.
8 Ibid.
9 Ibid., p. 490.

models with respect to a common research program. However, one crucial difference between analogies and mathematical idealizations is that the former can strongly influence the reception and the communication of scientific achievements to non-scientific and cross-disciplinary scientific audiences. Given my view that the making of research programs takes place in many different contexts of scientific practice, including science popularization and research management, this feature of analogies becomes especially important if one is interested in the expansion and historical continuity of research programs.

Other topics to which my understanding of the construction of analogy-based research programs might contribute, but for which a detailed argument has to be developed elsewhere, include social influences on scientific research practice and especially the impact of social structures on modeling. The Caltech case (*chapter 4*) can be used to show that scientific modeling takes place in specific social contexts and that the consideration of these contexts is crucial when it comes to the goals, results and impact of modeling processes. There is, of course, a certain difficulty in grasping *the social* in the process of modeling. When does a certain epistemic activity become social and which are the social dynamics that influence this activity? Which could be the important social events or factors one should look at in the case of modeling? Are there such things as non-social activities? A first answer that can be given to these questions with regard to my previous discussion of Pincock's anchor-thesis is that what "social influence" means clearly depends on the kinds of models we are considering and on the groups of people participating in model construction, reception and reconstruction. The social influences on mathematical modeling are presumably different ones than the ones which are present in analogical modeling with respect to the different kinds of actors and communities involved. One question that should be considered in this regard is whether the reception of models by non-scientific communities affects the interpretation and usage of these models by the next generation of scientists. Or in other words: Which actors and communities are involved in the re-interpretation of e.g. a mathematical model and whose re-interpretation lasts in the heads of the scientific community? In the present study, I focused on this particular question for the case of analogical modeling. But there are of course other aspects that should also be considered if one is interested in a social theory of scientific modeling. One of these aspects is the role of institutions and institutional dynamics. James Griesemer examines the role of natural history museums in the construction and preservation of material models in biology and claims that these museums strongly contribute to the realization of material "models of theoretical scientific entities and processes" through "the institution's personnel, procedures, traditions, and programs, collections of field notes and specimens." [10] He further argues that these activities can sometimes help to "realize models of theoretical scientific entities and processes."[11] Griesemer continues that, if applicable to other cases of modeling and kinds of institutions, his findings suggest "a way in which institutional analysis might be relevant to philosophical interpretations of science"; namely to the extent that the "structure and strategy of model-building depends on work organization."[12] Understood in this sense, "social structure not only provides a context in which intellectual work is done, it may also determine the course and, to some extent, the content of that work."[13]

In consideration of the study of lock-and-key analogy usage at hand, Griesemer's claim can, without doubt, be applied to analogy-based modeling and to other kinds of scientific institutions, such as philanthropic organizations and research institutes. As shown in chapter 4, the social structure of Caltech's biochemistry group and in particular its relationship to the Rockefeller Foundation clearly determined the course of biochemical research at the institute.

---

10 Griesemer (1990): Modeling in the museum: On the role of remnant models in the work of Joseph Grinnell, in: Biology and Philosophy, 5, pp. 3-36., p. 30.
11 Ibid.
12 Ibid.
13 Ibid, p. 36.

Literature

ARCHIVAL SOURCES

Paul Ehrlich Collection, RAC, 650 Eh 89 (Rockefeller Archive Center)

Manuscripts and Speeches

Lecture by Ehrlich about Salvarsan therapy, unknown occasion, November/December 1910, (Series I, Box 3, Folder 11).

Zettel Buch II und Carcinom, 1900 December 25 – 1901 September 21

"Kampf mit bordet", 14. Juni, 1901, (Box 8, Folder 2).
Notes to Hans Sachs and Julius Morgenroth related to the study of hemolysines, (Box 8, Folder 1 and 2).
"Schema für Hemmung", (Box 8, Folder 1).
"Bildung von Antistoffen", (Box 8, Folder 1).
Note to Morgenroth, (Box 8, Folder 1).
"Kampf mit bordet", 14. Juni, 1901, (Box 39).

Disputes and Polemics:

"Gruber Polemik", (Box 3, Folder 18).

Articles, Newsclippings and Manuscripts

Ehrlich Centennial: Annals of the New York Academy of Science (1954), (Box 63, Folder 6).

Invitation letter to the "Twenty-First Series Postgraduate Radio Programme of the New York Academy of Medicine", by the "Committee on Medical Information", August-October 1954, (Box 63, Folder 21).

Invitation letter to the "Dinner preceding the Celebration Meeting on Paul Ehrlich's 100th Anniversary at the New York Academy of Medicine, 2 East 103rd Street on Wednesday, March 10, 1954, (Box 63, Folder 21).

Commemoration of the 100th Anniversary of the Birth of Paul Ehrlich, March 10, 1954, (Box 63, Folder 6).

Pinkus M.D., Herman, "Paul Ehrlich and his Impact on Dermato-Syphilology" t.s., undated, (Box 64, Folder 20).

Witebsky, E. (1954): Side chain theory, in: Ehrlich Centennial, Annals of the New York Academy of Science, New York, (Box 62, Folder 6).

Bauer (1954): Paul Ehrlich's influence on chemistry and biochemistry, in: Ehrlich Cent., Annals of the New York Academy of Science, p. 151 (Box 62, Folder 6).

Clippings of the movie "Dr. Ehrlich's magic bullets."

Letter from Edward G. Robinson to Hedwig Ehrlich, March 22, 1940, (Folder 18).

The movie's original script (Box 63, Folder 18)

Rockefeller Foundation Records (Rockefeller Archive Center)

RG 1.1: Projects: Medical, Health, and Population Sciences, Series 205, The California Institute of Technology, Immunology and Immunochemistry

"A proposed project of experimental investigation of the structure of antibodies and the nature of immunological reactions", March 18, 1941. (Box 7, Folder 92).

Albert Tyler's report "Of the work supported in part by a grant from the Rockefeller Foundation during the year starting July 1, 1939" (Box 6, Folder 83)

Letter from Hanson to Tyler, December 1942 (Box 7, Folder 93).

Letter from Hanson to Pauling, November 2nd 1944 (Box 7, Folder 95).

Recommendation letter from Beadle to Hanson, May 24, 1943 (Box 7, Folder 94).
Recommendation letter from Frank R. Lillie to Hanson, May 20, 1943 (Box 7, Folder 94).
"Report on work supported in part by a grant from the Rockefeller Foundation during the year starting July 1, 1939" (Box 6, Folder 83).
Letter from Roger Adams to Weaver, April 6, 1938 (Box 6, Folder 79).
Weaver's response to Roger Adams, April 13, 1938 (Box 6, Folder 79).
Letter from Pauling to Weaver, August 11, 1939. (Box 6, Folder 82).

RG 1.2: Projects: Agricultural and Natural Sciences, Series 205, The California Institute of Technology

Weaver's manuscript of "Why should Joe Willits care about the structure of proteins?", Inter-Office Correspondence, September 20, 1951 (Box 4, Folder 27).
Rockefeller Foundation: Proposed Program of Research; Chemistry and Biology Progress Report 1946-47 (Box 4, Folder 22).
"A proposed program of research on the fundamental problems of Biology and Medicine" by the Division of Chemical Engineering of Caltech [for submission to RF and the National Foundation of Infantile Paralysis], May 17, 1946 (Box 4, Folder 22).

RG 3.1: Administration, Program and Policy Records (1910-2000), Series 915

The Rockefeller Foundation: "The natural and medical sciences cooperative program", December 13, 1933 (Box 1, Folder 7).

Warren Weaver Papers, Collections of Individuals (Rockefeller Archive Center)

Letter from Weaver to Pauling, December 19, 1933, (Box 10, Folder 136).
Letter from Pauling to Weaver, November 22, 1934, (Box 10, Folder 136.
Letter from Weaver to Pauling, November 23, 1934, (Box 10, Folder 136).
Letter Weaver to Pauling, October 3, 1934 (Box 6, Folder 73).
Diary note from November 6, 1937 (Warren Weaver Diaries, Box 6, Folder 78).

Announcement from the "Office of Public Relations" concerning the joint project in Biology and Chemistry, April 29, 1948, RG 1.2, Series 205, Box 4, Folder 25.

Biology Divisional Records (Caltech Archives)

Genetics papers (Box 4, Folder 11).
Chemistry-Biology-Program, 1947-51, Church Laboratories 1952-53, (Box 22, Folder 11, 12, 19).
Medical Research Chemistry, Laboratory of 1950; Fund for Chemical Genetics, 1946-53, (Box 47, Folder 11, 18).
National Foundation for Infantile Paralysis, Reports (1949; 1950/51), (Box 49, Folder 15, 16).
National Society for Medical Research (1948), (Box 55, Folder 5).
President's Office, since 1954; President's Report (July 1, 1947-50), (Box 60, Folder 2, 3).
Research Program Proposal, 1945; Reports and Manuscripts 1949, (Box 62, Folder 13, 24).
Reports and Manuscripts, 1950 – June 1951; July 1951-1954; Chemical Biology Grant 1953; Rockefeller Foundation Fund 1949-50; Rockefeller Foundation Proposed Program of Research, Chemistry and Biology Progress Report 1946-47, (Box 63, Folder 1, 2, 3, 4, 6).
Tyler, Albert 1945-54, (Box 69, Folder 4).

Ava-Helen and Linus Pauling Papers (Special Collections and Archives of the Oregon State University)

03 Manuscripts and Typescripts of Articles by Linus Pauling

Partial Manuscript, Typescript, Referee's Comments, Figures: "A Theory of the Structure and Process of Formation of Antibodies", July 27, 1940; Folder 5: Typescript: "The Nature of the Intermolecular Forces Operative in Biological Processes", 1940, (Box 1940a, Folder 2).
Manuscript, Typescript: "Properties of Antibodies", September 26, 1941, (Box 1941a, Folder 1).

Typescript, Correspondence: "Molecular Structure and Intermolecular Forces", August 1942; Folder 7: Manuscript, Typescript: "The Production of Antibodies in vitro", 1942, (Box 1942a, Folder 6).
Typescript: "The Division of Chemistry and Chemical Engineering at the California Institute of Technology", August 15, 1944, (Box 1944a, Folder 5).
Typescript: "A Proposed Program of Research on the Fundamental Problems of Biology and Medicine", February 14, 1946, (Box 1946a, Folder 1).
Typescript, Correspondence: "Molecular Structure and Biological Specificity", July 17, 1947; Folder 5: Correspondence: "The Structure of Antibodies and the Nature of Specific Biological Forces", December 5, 1947; Folder 7: Galley Proof: "Chemical Achievement and Hope for the Future", 1947, (Box 1947a, Folder 3).
Typescript, Abstract: "The Structure of Antibodies and the Nature of Serological Reactions", April 8, 1948, (Box 1948a, Folder 3).

04 Manuscripts and Typescripts of Speeches by Linus Pauling

Manuscript Notes: "The Nature of Serological Reactions - The Structure and Formation of Antibodies", Chemistry Seminar, University of California, Berkeley, March 12, 1940, (Series 1, Box 1940).
Manuscript, Program: "Development of Immunochemistry at CIT", Alumni Seminar, California Institute of Technology, Pasadena, April 11, 1943; Series 2: Manuscript: "Chemical Studies of the Structure of Antibodies", Chemical Society of Washington and Washington Academy of Sciences, Washington, D.C., April 22, 1943; Series 7: Program, Notes, Ancillary Material: "Specificity of Intermolecular Interaction", American Physical Society, Pasadena, California, December 27, 1943, (Series 1, Box 1943).
Manuscript Notes: "The Future of Medical Research", Talk for UMCA, California Institute of Technology, Pasadena, August 29, 1945; Series 5: Manuscript: "The Future of Scientific and Medical Research", Pasadena, California, September 26, 1945, (Series 3, Box 1945).
Manuscript: "Structure and Specificity", Eli Lilly and Company, No Location, March 19, 1946; Series 6: Manuscript: "Medical Research of the Future", Third William J. Stone Memorial Lecture, Pasadena, Califor-

nia, April 2, 1946; Series 7: Manuscript, Typescript, Program, Notes, Ancillary Material: "Molecular Architecture and Biological Reactions", The George Westinghouse Centennial Forum, Pittsburgh, Pennsylvania, May 17, 1946; Series 11: 1946s: Transcript, Correspondence, Ancillary Material: "Molecular Architecture and Medical Progress", United States Rubber Company New York Philharmonic-Symphony Intermission Feature, Columbia Broadcasting System, New York, October 13, 1946, (Series 5, Box 1946).

Correspondence, Flyer: "The Future of Scientific and Medical Research", Annual Regional Dinner Meeting of the American College of Physicians, Southern California chapter, Los Angeles, California, February 7, 1947; Series 15: Typescript, Correspondence, Reprint, Background Material: "Molecular Structure and Biological Specificity" and "The Nature of Bonds in Metals and Intermetallic Compounds", Congress Lecturer, 11th International Congress of Pure and Applied Chemistry, London, England, July 17-24, 1947; Series 25: Correspondence, Itinerary: "Intermolecular Forces and Molecular Structure in Relation to Biological Specificity", E. I. DuPont de Nemours & Co., Wilmington, Delaware, December 23, 1947, (Series 1, Box 1947).

Manuscript: "Intermolecular Forces and Biological Specificity", Oxford Lectures, Lecture 1, England, April 27, 1948; Correspondence, Flyer: "Molecular Architecture and the Processes of Life", The Sir Jesse Boot Lectureship, University College, Nottingham, England, May 28, 1948, (Series 32, Box 1948).

PRIMARY SOURCES (PUBLISHED)

Abderhalden, Emil (1919): Die Bedeutung von Fischers Lebenswerk für die Physiologie, in: Die Naturwissenschaften, 7 (46), pp. 860-868.

Appolant, H. et al. (1914): Paul Ehrlich. Darstellung seines wissenschaftlichen Wirkens. Festschrift zum 60. Geburtstag des Forschers, Jena.

Bäumler, Ernst (1971): Auf der Suche nach der Zauberkugel – Vom grossen Abenteuer der modernen Arzneimittelforschung, Düsseldorf.

Bäumler, Ernst (1979): Paul Ehrlich – Forscher für das Leben, Möchengladbach.

Bäumler, Ernst (1988): Die Rotfabriker, München.

Bäumler, Ernst (1989): Farben, Formeln, Forscher, Berlin.

Bechhold, Heinrich (1905): Ungelöste Fragen über den Anteil der Kolloidchemie an der Immunitätsforschung, in: Wiener klinische Wochenschrift, 18.

Breinl, Friedrich/Haurowitz, Felix (1930): Untersuchung des Präzipitates aus Hämoglobin und Antihämoglobin-Serum und einige Bemerkungen zur Natur der Antikörper, in: Zeitschrift für Physiologische Chemie, 192, pp. 45-57.

Dieterle, William (director, 1940): Dr. Ehrlich's Magic Bullet, USA.

Ehrlich Centennial (1954): Annals of the New York Academy of Science.

Ehrlich, Paul (1885): Das Sauerstoff-Bedürfniss des Organismus. Eine Farbanalytische Studie, Berlin. For the english translation of the title see Dale, Henry H.: Introduction, in: Himmelweit, F. (1956): The Collected Papers of Paul Ehrlich, London and New York.

Ehrlich, Paul (1900): On immunity with special reference to cell life, in: Proceedings of the Royal Society of London, 66, pp. 424-448.

Ehrlich, Paul (1904): Gesammelte Arbeiten zur Immunitätsforschung, Frankfurt am Main.

Ehrlich, Paul (1909): Beiträge zur experimentellen Pathologie und Chemotherapie, Bd. 1, Leipzig.

Ehrlich, Paul (1909): Chemotherapeutische Trypanosomstudien, in: Beiträge zur experimentellen Pathologie und Chemotherapie, pp. 97-115.

Ehrlich, Paul (1960 {1906}): Address delivered at the dedication of the Georg-Speyer-Haus (September 6, 1906), in: Himmelweit, F. (ed.): The collected papers of Paul Ehrlich, Third Volume (Chemotherapy), London, pp. 53-63.

Ehrlich, Paul (1960 {1907}): Chemotherapeutische Trypanosomen Studien, in: Himmelweit, F. (ed.) (1960): Paul Ehrlich, Gesammelte Arbeiten, Dritter Band, London/New York/Paris, pp. 81-106, first published in 1907 in: Berliner Klinische Wochenzeitschrift, 44, pp. 233-236, 280-283, 310-314, and 341-344.

Ehrlich, Paul (1993 {1878}): Contributions to the theory and practice of histological staining, Inaugural Dissertation, University of Leipzig, in: Himmelweit, F. (ed.): The collected papers of Paul Ehrlich in four volumes, Münster.

Ehrlich, Paul/Morgenroth, Julius (1904 {1889}): Ueber Haemolysine I-V, in: Gesammelte Arbeiten zur Immunitätsforschung, pp. 16-105.

Ehrlich, Paul/Morgenroth, Julius (1904 {1899}): Zur Theorie der Lysinwirkung, in: Gesammelte Arbeiten zur Immunitätsforschung, pp. 1-16.
Ehrlich, Paul/Morgenroth, Julius (1904 {1899}): Ueber Haemolysine, 4. Mittheilung, in: Ehrlich (1904): Gesammelte Arbeiten zur Immunitätsforschung, Berlin, pp. 73-86.
Ehrlich, Paul/Morgenroth, Julius (1904 {1899}): Ueber Haemolysine, 2. Mittheilung, in: Gesammelte Arbeiten, pp. 16-35.
Ehrlich, Paul/Morgenroth, Julius (1904 {1899}): Ueber Haemolysine, 5. Mittheilung, Gesammelte Arbeiten, pp. 86-110.
Ehrlich, Paul/Morgenroth, Julius (1904 {1899}): Zur Theorie der Lysinwirkung, in: Gesammelte Arbeiten (1904), pp. 1-15.
Fischer, Emil (1890): Synthesen in der Zuckergruppe I, in: Berichte der Deutschen Chemischen Gesellschaft (23), pp. 2114-2141. Also published in: Fischer (1909): Untersuchungen über Kohlenhydrate, pp. 18-20.
Fischer, Emil (1890): Über die optischen Isomeren des Traubenzuckers, der Gluconsäure und der Zuckersäure, in: Fischer (1909): Untersuchungen über Kohlenhydrate und Fermente, pp. 362-377.
Fischer, Emil (1891): Ueber die Konfiguration des Traubenzuckers und seiner Isomeren, in: Berichte (24), pp. 1836-1845. Also published in: Fischer (1909): Untersuchungen über Kohlenhydrate und Fermente, pp. 417-427.
Fischer, Emil (1891): Ueber die Konfiguration des Traubenzuckers und seiner Isomere II, in Berichte (24), pp. 2683-2687.
Fischer, Emil (1894): Die Chemie der Kohlenhydrate und ihre Bedeutung für die Physiologie. Rede gehalten zur Feier des Stiftungstages der Militärärztlichen Bildungsanstalten am 2. August 1894, in: Id. (1909): Untersuchungen über Kohlenhydrate, pp. 96-115.
Fischer, Emil (1894): Einfluss der Configuration auf die Wirkung der Enzyme I, in: Berichte der Deutschen Chemischen Gesellschaft, Vol. 27 (1894), pp. 2985-2993. Also published in: Fischer (1909): Gesammelte Werke: Untersuchung über Kohlenhydrate und Fermente II, ed. Max Bergmann, Berlin/Heidelberg, pp. 836-844.
Fischer, Emil (1894): Einfluss der Configuration auf die Wirkung der Enzyme II, in: Berichte (27), pp. 3479-3483.
Fischer, Emil (1894): Einfluss der Configuration auf die Wirkung der Enzyme, in: Berichte (27), pp. 2985-2993.

Fischer, Emil (1894): Synthesen in der Zuckergruppe II, in Berichte (27), pp. 3189-3232.
Fischer, Emil (1898): Bedeutung der Stereochemie für die Physiologie, in: Berichte der deutschen chemischen Gesellschaft (26), pp. 60-87.
Fischer, Emil (1909 {1888}): Über die Verbindung des Phenylhydrazins mit den Zuckerarten I-V, in: Untersuchungen über Kohlenhydrate, pp. 138-176.
Fischer, Emil (1909 {1894}): Die Chemie der Kohlenhydrate und ihre Bedeutung für die Physiologie, in: Untersuchungen über Kohlenhydrate und Fermente, pp. 96-115.
Fischer, Emil (1909 {1898}): Bedeutung der Stereochemie für die Physiologie, in: Untersuchungen über Kohlenhydrate und Fermente, pp. 116-137.
Fischer, Emil (1909 {1899}): Über die Spaltung racemischer Verbindungen in die activen Componenten, in: Untersuchungen über Kohlenhydrate und Fermente, pp. 890-892.
Fischer, Emil (1923 {1907}): Proteine und Polypeptide, in: Fischer/Bergman (eds.) (1923): Untersuchungen über Proteine, Polypeptide und Aminosäuren, Berlin, pp. 748-757.
Fischer, Emil (1924 {1907}): Organische Synthese und Biologie (Faraday Lecture), in: Untersuchungen aus verschiedenen Gebieten, p. 778-795.
Fischer, Emil/Thierfelder, Hans (1894): Verhalten der verschiedenen Zucker gegen reine Hefen, in: Berichte (27), pp. 2031-2037.
Fischer, Emil/Thierfelder, Hans (1909 {1894}): Verhalten der verschiedenen Zucker gegen reine Hefe, in: Untersuchungen über Kohlenhydrate und Fermente, p. 829-835.
Haurowitz, Felix (1947): Antibodies: Their Nature and Formation, in: The Lancet (Special Articles), Jan. 25, pp. 149-151.
Heidelberger, Michael (1939): Quantitative absolute method in the study of antigen-antibody reactions, in: Bacteriological Review, 3 (1), pp. 49-95.
Hoechst, Abteilung für Öffentlichkeitsarbeit (1980): Paul Ehrlich: Forscher für das Leben. Reden zum 125. Geburtstag des Forschers.
Meyer, Viktor (1890): Ergebnisse und Ziele der stereochemischen Forschung, in: Berichte der deutschen chemischen Gesellschaft (23), pp. 567-619.

Michaelis, Leonor (1908): Physikalische Chemie der Kolloide, in: Korànyi/Richter (eds.): Physikalische Chemie und Medizin: Ein Handbuch, Leipzig, pp. 341-453.

Morgenroth, Julius (1905): Ueber die Wiedergewinnung von Toxin aus seiner Antitoxinverbindung, in: Berliner klinische Wochenzeitschrift, 42, pp. 1550-1554.

Mudd, Stuart (1932): A hypothetical mechanism of antibody formation, in: Journal of Immunology, 23, pp. 423-427.

Osborn, Thomas W. B. (1937): Complement or Alexin, London.

Pasteur, Louis (1860): Ueber die Asymmetrie bei natürlich vorkommenden organischen Verbindungen, 2 Vorträge gehalten am 20. Januar und 3. Februar 1860 in der Société chimique zu Paris, translated and edited by Ladenburg, Leipzig.

Pauling, Linus (1940): A theory of the structure and process of antibody formation, in: Journal of the American Chemical Society, 62 (10), pp. 2643-2657.

Pauling, Linus/Delbrück, Max (1940): The nature of the intermolecular forces operative in biological processes, in: Science, 92 (2378), pp. 77-79.

Pauling, Linus/Campbell, Dan/Pressman, David (1943): The Nature of the Forces between Antigen and Antibody and of the Precipitation Reaction, in: Physiological Reviews, 23 (3), pp. 203-219.

Pauling, Linus (1945): Molecular structure and intermolecular forces, in: Landsteiner (ed.): The Specificity of Serological Reactions, Cambridge (Ma), pp. 275-293.

Pauling, Linus (1946): Molecular Architecture and Biological Reactions, in: Biological Science, 24 (10), pp. 1375-1377.

Pauling, Linus (1946): Molecular Architecture and Medical Progress, Radio talk broadcast on the New York Philharmonic-Symphony Radio Program sponsored by the U. S. Rubber Co., October 13, New York.

Pauling, Linus (1948): Chemical achievement and hope for the future, Silliman Lecture presented at Yale University in October, 1947, on the occasion of the Centennial of the Sheffield Scientific School, in: American Scientist, 36, pp. 51-58.

Pauling, Linus (1948): Molecular Architecture and the Processes of Life, 21st Sir Jesse Boot Foundation Lecture, May 28, 1948, Nottingham, England, pp. 1-13.

Pauling, Linus (1948): Molecular structure and biological specificity, Presidential address at Section 2, 11th International Congress of Pure and Applied Chemistry, London, July 17-24 (1947), in: Chem. Ind. Supple, pp. 1-4. 86-109.

Pauling, Linus (1948): The nature and forces between large molecules of biological interest, Friday evening discourse at the Royal Institution of Great Britain, London, on February 27, in: Nature (161), pp. 707-709.

Pauling (1949): Structural Chemistry in Relation to Biology and Medicine, in: Baskerville Chemical Journal, 1 (1), pp. 4-7.

Pauling, Linus/Itano, Harvey/Singer, S. J. (1949): Sickle cell anemia, a molecular disease, in: Science, 110, pp. 543-548.

Pauling, Linus (1975): The molecular basis of biological specificity, in: Nature, 248, pp. 769-771, p. 769.

Pinkus, Hermann (1954): In Commemoration of the 100th Anniversary of the Birth of Paul Ehrlich, Reprinted from the American Journal of Clinical Pathology, 24 (7), July 1954, Received for publication March 19, 1954.

Tyler, Albert (1940): Sperm agglutination in the keyhole limpet, in: The Biological Bulletin, LXXVIII (2), pp. 159-178.

Tyler, Albert (1948): Fertilization and Immunity, in: Physiological Review, 28 (2), pp. 180-219.

Van't Hoff, Jacobus H. (1875): La chimie dans l'espace, Rotterdam.

Van't Hoff, Jacobus H. (1877): Lagerung der Atome im Raume, Braunschweig.

Wislicenus, Johannes (1888): Ueber die räumliche Anordnung der Atome in organischen Molekülen und ihre Bestimmung in geometrisch ungesättigten Verbindungen (in zwei Bänden), in: Akademie der Wissenschaften. Abhandlungen (Math. Physik), Band 14, Leipzig.

Witebsky, E. (1954): Side chain theory, in: Ehrlich Centennial, Annals of the New York Academy of Science (1954).

SECONDARY SOURCES

Abrantes, Paolo (1999): Analogical reasoning and modeling in the sciences, Foundations of Science, 4 (3), pp. 237-240.

Alexander, Patricia A./White, Stephen C./Daugherty, Martha (1997): Analogical reasoning and early mathematical learning, in: English (ed.): Mathematical reasoning: Analogies, metaphors, and images, pp. 117-147.

Ankeny, Rachel/Leonelli, Sabina (2016): Repertoires: a post-Kuhnian perspective on scientific change and collaborative research, in: Studies in History and Philosophy of Science Part A, 60, pp. 18-28.

Authier, André (2013): Early Days of X-ray Crystallography, Oxford.

Bailer-Jones, Daniela (2000): Scientific Models as Metaphors, in: Hallyn (ed.): Metaphor and Analogy in the Sciences, Dordrecht, pp. 181-198.

Bailer-Jones, Daniela (2008): Models, Metaphors, and Analogies, in: The Blackwell Guide to Philosophy of Science, pp. 108-127.

Barnett, James A./Lichtenthaler, Frieder W. (2001): A history of research on yeast 3: Emil Fischer, Eduard Buchner and their contemporaries, 1880-1900, in: Yeast, 18, pp. 363-388.

Barwich, Anne-Sophie (2013): Making Sense of Smell: Classification and Model Thinking in Olfaction Theory, Doctoral thesis, University of Exeter.

Bechtel, William/Richardson, Robert (2010 {1993}): Discovering Complexity. Decomposition and Localization as Strategies of Scientific Research, Cambridge (Ma).

Black, Max (1962): Models and Metaphors, Cornell.

Black, Max (1977): More about Metaphor, in: Ortony (ed.): Metaphor and Thought, Cambridge.

Boumans, Marcel (1999): Built-in justification, in: Morgan and Morrison (eds.): Models as Mediators, Cambridge (Ma), pp. 66-96.

Brock, William H. (1997): Viewegs Geschichte der Chemie, Braunschweig/ Wiesbaden.

Budd, Robert (2007): Penicillin: Triumph and Tragedy, Oxford.

Butterfield, James (1968 {1931}): The Whig Interpretation of History, London.

Cambrosio, Alberto/Jacobi, Daniel/Keating, Robert (2005): Arguing with images: Pauling's Theory of Antibody formation, in: Representations, 89 (1), pp. 94-130.

Cambrosio, Alberto/Jacobi, Daniel/Keating, Peter (2004): Intertextualité et archi-iconicité: le cas des représentations scientifiques de la réaction antigène-anticorps, in: Études de communication, 27, pp. 2-13.

Cambrosio, Alberto/Keating, Peter/Jacobi, Daniel (1993): Ehrlich's "beautiful pictures" and the controversial beginnings of immunological imagery, in: Isis, 84 (4), pp. 662-692.

Carrier, Martin (2002): Explaining Scientific Progress: Lakatos' Methodological Account of Kuhnian Patterns of Theory Change, in: Kampis/Kvasz/Stölzner (eds.): Appraising Lakatos: Mathematics, Methodology, and the Man, Vol. 1 of the Series Vienna Circle Institute Library, pp. 53-71.

Cartwright, Nancy (1997): Models: The blueprints for laws, in: Philosophy of Science 64 (4), pp. S292-S303.

Chabner, Bruce A./Roberts, Thomas G. Jr. (2005): Chemotherapy and the war on cancer, in: Nature Reviews, 5, pp. 65-72.

Christie, Robert (2001): Color Chemistry, London.Cruse, Julius M./Lewis, Robert E. (eds.) (2010): Atlas of Immunology, Third Edition.

Del Re, Guiseppe (2000): Models and analogies in science, in: HYLE, 6 (1), pp. 5-15.

Eisenbach, Ulrich (2010): Speyer, in: Neue Deutsche Biographie (NDB), Band 24, Duncker & Humblot, Berlin, pp. 674-676.

Finsham, John R.S. (1994): Sterling Emerson, 1900-1988. A biographical memoir, published by the National Academy of Sciences, Washington DC.

Fleck, Ludwik (1935): Genesis and development of a scientific fact, Chicago.

Francoeur, Eric (2000): Beyond dematerialization and inscription, in: HYLE – International Journal for Philosophy of Chemistry, 6 (1), pp. 63-84.

Frigg, Roman and Hartmann, Stefan (2012): Models in Science, in: The Stanford Encyclopedia of Philosophy: http://plato.stanford.edu/archives/fall2012/entries/models-science/, 10/08/2015, 14:00.

Fruton, Joseph (1985): Contrasts in Scientific Style. Emil Fischer and Franz Hofmeister: Their Research Groups and Their Theory of Protein Structure, in: Proceedings of the American Philosophical Society, 129 (4), pp. 313-370.

Fruton, Joseph (2002): Methods and Styles in the development of chemistry, in: Memoirs of the American Philosophical Society held at Philadelphia for promoting useful knowledge, Vol. 245, Philadelphia.

Gaboa, Steven (2008): In defense of analogical reasoning, in: Informal Logic, 28 (3), pp. 229-241.

Gelfert, Alexander (2011): A model-based representation in scientific practice, in: History and Philosophy of Science Part A, 42 (2), pp. 251-252.

Gilbert, Scott F. and Greenberg, Jason P. (1984): Intellectual Traditions in The Life Sciences. II. Stereocomplementarity, in: Perspectives in Biology and Medicine, 28 (1), pp. 18-34.

Glaesmer, Röderich (2004): Zur Entwicklung der wissenschaftlichen Verflechtung der Chemie mit anderen Wissenschaften bei der Erforschung von Struktur, Funktion und Synthese von Proteinen im 20. Jahrhundert, Dissertation, Berlin.

Godfrey-Smith, Peter (2006): The strategy of model-based science, in: Biology and Philosophy, 21, pp. 725-740.

Hager, Thomas (1995): Force of nature: The life of Linus Pauling, Michigan.

Hager, Thomas (1998): Linus Pauling and the chemistry of life, Oxford.

Hargittai, M./Hargittai, I. (2008): Symmetry through the eyes of a chemist, Heidelberg.

Hesse, Mary (1966): Models and Analogies in Science, Notre Dame.

Hoesch (1921): Emil Fischer. Sein Leben und sein Werk, in: Berichte der deutschen chemischen Gesellschaft, Sonderheft des 54. Jahrgangs.

Hofstadter, Douglas (1995): Fluid Concepts and Creative Analogies, New York.

Holmes, Larry (2004): Patterns and stages in the careers of experimental scientists, New Haven.

Hudson, Claude (1941): Emil Fischer's Discovery of the Configuration of Glucose, in: Journal of chemical education, pp. 353-357.

Hudson, Claude (1948): Historical Aspects of Emil Fischer's Fundamental Conventions for Writing Stereo-Formulas in a Plane, in: Advances in Carbohydrate Chemistry, 3, pp. 1-22.

Hüntelmann, Axel (2010): Legend of science: External constructions by the extended family –the biography of Paul Ehrlich, in: InterDisciplines 2, pp. 13-36.

Hüntelmann, Axel (2011): Paul Ehrlich. Leben, Forschung, Ökonomien, Netzwerke, Göttingen.

Hüntelmann, Axel (2013): Making Salvarsan. Experimental Therapy and the Development and Marketing of Salvarsan at the Interface between Science, Clinic, Industry and Public Health, in: Gaudillière/Hess (eds.): Ways of Regulating Drugs in the 19th and 20th Centuries, Basingstoke, pp. 43-65.

James, Jeremiah (2014): Modeling the scale of atoms and bonds: The origin of space-filling parameters, in: Klein/Reinhardt (eds.): Objects of chemical inquiry, Sagamore Beach, pp. 281-320.

Kasten, Frederick H. (1996): Paul Ehrlich. Pathfinder in Cell Biology, in: Biotechnology and History of Chemistry, 71 (1), pp. 2-37.

Kay, Lily E. (1989): Molecular Biology and Pauling's Immunochemistry: A Neglected Dimension, in: History and Philosophy of the Life Sciences, 11 (2), pp. 211-219.

Kay, Lily E. (1993): The Molecular Vision of Life: Caltech, The Rockefeller Foundation, and the Rise of the New Biology, Oxford.

Kay, Lily E. (2000): Who wrote the book of life? A History of the Genetic Code, Stanford.

Klein, Ursula (2003): Experiments, models, paper tools, Stanford.

Klein, Ursula (2001): The Creative Power of Paper Tools in Early Nineteenth Century, in: Klein, Ursula (ed.): Tools and Modes of Representation in the Laboratory Sciences, Dordrecht.

Knuuttila, Tarja/Loetgers, Andrea (2014): Varieties of noise: Analogical reasoning in synthetic biology, in: Studies in History and Philosophy of Science, Part A, 48, pp. 76-88.

Knuuttilla, Tarja/Voutilainen, Atro (2003): A parser as an epistemic artifact: A material view on models, in: Philosophy of Science, 70 (5), pp. 1484-1495.

Kohler, Robert (1991): Partners in science: Foundations and Natural Scientists, 1900-1945, Chicago.

Kuhn, Thomas S. (1970 {1962}): The Structure of Scientific Revolutions, Chicago.

Lakatos, Imre (1970): Falsification and the Methodology of Scientific Research Programs, in: Lakatos, Imre/Musgrave, Alan (eds.): Criticism and the Growth of Knowledge.

Laszlo, Pierre (1986): Molecular correlates of biological concepts, in: Comprehensive Biochemistry, 34 A.

Laszlo, Pierre (1999): Circulation of concepts, in: Foundations of chemistry, 1 (3), pp. 225-238.

Lausen, Fabian (2014): Zur heuristischen Qualität des Reduktionismus, Münster.

Lederer, Susan E./Parascandola, John (1998): Screening Syphilis: Dr. Ehrlich's Magic Bullet Meets the Public Health Service, in: Journal of the History of the Medical Allied Sciences, 53 (4), pp. 345-70.

Lenoir, Timothy (1988): A magic bullet: Research for profit and growth of knowledge in Germany around 1900, in: Minerva 26 (1), pp. 66-88.

Lesch, John E. (2007): The first miracle drug: How the sulfa drugs transformed medicine, Oxford.

Lichtenthaler, Frieder W. (1994): Hundert Jahre Schlüssel Schloss Prinzip: Was führte Emil Fischer zu dieser Analogie? In: Angewandte Chemie 106, pp. 2456-2467.

Löwy, Ilana (1990): The strength of loose concepts: The case of immunology, in: History of Science, 30 (90), pp. 371-396.

Maehle, Andreas H. (2004): Receptive substances, in: Medical History, 48, pp. 153-174.

Maehle, Andreas H./Prüll, Cay-Rüdiger/Halliwell, Robert F. (2002): Emergence of drug receptor theory, in: Nature Reviews Drug Discovery, 1, pp. 637-641.

Magnani, Lorenzo/Nersessian, Nancy/Thagard, Paul (1999): Model-based reasoning in scientific discovery, Dordrecht.

Martin, Emily (1991): The egg and the sperm: How science has constructed a romance based on stereotypical male-female roles, in: Signs, 16 (3), pp. 485-501.

Mazumdar, Pauline (1989): The template theory of antibody formation and the chemical synthesis of the twenties, in: Mazumdar, Pauline (ed.): Immunology 1930-1980 –Essays on the history of immunology, Toronto, pp. 13-32.

Mazumdar, Pauline (1995): Species and Specificity. An interpretation of the history of immunology, Cambridge.

Meinel, Christoph (2004): Molecules and croquet balls, in: Chadarevian/ Hopwood (eds.): Models: The Third Dimension of Science, pp. 242-276.
Meinel, Christoph (2009): Kugel und Stäbchen. Vom Kulturellen Ursprung chemischer Modelle, in: Kultur und Technik, 33 (2), pp. 14-21.
Morange, Michel (1998): A History of Molecular Biology, Cambridge (Ma).
Morange, Michel (2010): What history tells us XX. Felix Haurowitz (1896-1987) –A difficult journey in the political and scientific upheavals of the 20th century, in: Journal of the Biosciences, 35 (1), pp. 17-20.
Morgan, Mary (2012): The World in the Model: How Economists Work and Think, Cambridge.
Morgan, Mary/Morrison, Margaret (1999): Models as mediating instruments, in: Morgan, Mary/Morrison, Margaret (ed.): Models as mediators, Cambridge (Ma).
Moulin, Anne Marie/Harshav, Barbara (1988): Text and Context in Biology: In Pursuit of the Chimera, in: Poetics Today, 9 (1), pp. 145-161.
Mund, Marianne (1999): Struktur, Konfiguration und Formelschreibweise der Kohlenhydrate von Kekulés und Coupers Valenzlehre (1858) bis zum Beginn von Emil Fischers Arbeiten über die Kohlenhydrate (1890), Hamburg/Bukarest.
Musgrave (1976): Method or Madness? Can the methodology of research programmes be rescued from epistemological anarchism, in: Musgrave (ed.): Essays in Memory of Imre Lakatos, Boston Studies in the Philosophy of Science, 39, pp. 457-491.
Nersessian, Nancy (1988): Reasoning from imagery and analogy in scientific concept formation, in: PSA: Proceedings of the Biennial Meeting of the Philosophy of Science Association, pp. 41-47.
Nersessian, Nancy J. (1999): Model-based reasoning in conceptual change, in: Magnani, Nersessian and Thagard (eds.): Model-based reasoning in scientific discovery, New York.
Nersessian, Nancy J. (2002): Model-based reasoning: Science, Technology and Values, Dordrecht.
Nersessian, Nancy J. (2008): Creating Scientific Concepts, Cambridge (Ma).

Nersessian, Nancy J. (2009): How do engineering scientists think? Model-based simulation in biomedical engineering research laboratories, in: Topics in Cognitive Science, 1, pp. 730-757.

Nersessian, Nancy J./McLeod (2017): Interdisciplinary problem solving. Emerging models in Integrative Systems Biology, in: European Journal of Philosophy of Science (fc, in press).

Nersessian, Nancy J./Osbeck (2017): Epistemic Identities in Interdisciplinary Science, in: Perspectives on Science 25 (fc).

Nye, Mary Joe (1993): From chemical philosophy to theoretical chemistry: Dynamics of matter and dynamics of disciplines, 1800-1950, Berkeley/London.

Nye, Mary Joe (1999): Before big science: The pursuit of modern chemistry and physics, 1800-1940, Cambridge (Ma).

Nye, Mary Joe (2000): From student to teacher: Linus Pauling and the reformulation of the principles of chemistry in the 1930s, in: Lundgren/Bensaude-Vincent (eds.): Communicating chemistry: Textbooks and their audiences, 1789-1939, European Studies in Science History and the Arts, 3, pp. 397-414.

Nye, Mary Joe (2000): Physical and biological modes of thought in the chemistry of Linus Pauling, in: Studies in the history and philosophy of science, 31, pp. 475-491.

Nye, Mary Joe (2001): Paper tools and molecular architecture in the chemistry of Linus Pauling, in: Klein (ed.): Tools and modes of representation in the laboratory sciences, Boston, pp. 117-132.

Olby, Robert (1990): The molecular revolution in biology, in: Deby et al. (eds.): Companion to the history of modern science, London, pp. 503-520.

Paradowsky, Robert (2010): "Pauling Chronology". The Ava Helen and Linus Pauling Papers (online), OSU Special Collections & Archives. https://paulingblog.wordpress.com/2009/03/10/paradowskis-pauling-chronology/, 09/27/2018., 21:14.

Parascandola, John/Jasensky, Robert (1974): Origins of the receptor theory of drug action, in: Bulletin of the History of Medicine, 48, pp. 199-220.

Pincock, Christopher (2012): Mathematical models of biological patterns: Lessons from Hamilton's selfish herd, in: Biological Philosophy, 27, pp. 481-496.

Prüll, Cay-Rüdiger (2003): Part of a scientific master plan? Paul Ehrlich and the origins of his receptor concept, in: Medical History, 47, pp. 332-356.

Ramberg, Peter (2003): Chemical Structure, Spatial Arrangement. The early history of stereochemistry, 1874-1914, Burlington.

Ramsay, Peter (1975): Van't Hoff-Le Bel Centennial, ACS Symposium Series, Washington, DC.

Reinhardt, Carsten (2014): The olfactory object. Towards a history of smell in the 20th century, in: Klein/Reinhardt (eds.): Objects of Chemical Inquiry, Sagamore Beach, pp. 321-341.

Rheinberger, Hans-Jörg (1995): Kurze Geschichte der Molekularbiologie, in: Preprints of the Max Planck Institute for the History of Science (24), pp. 1-46.

Richardson, Robert (1999): Heuristics and satisficing, in: Bechtel/Graham: A companion to cognitive science, Blackwell.

Rocke, Alan (2010): Image and Reality. Kekulé, Kopp, and the Scientific Imagination, Chicago.

Sarasin, Philipp/Berger, Silvia/Hänseler, Marianne et al. (2006): Bakteriologie und Moderne. Eine Einführung, in: Ead. (eds.): Bakteriologie und Moderne: Studien zur Biopolitik des Unsichtbaren, pp. 8-43.

Schaffner, Kenneth (1993): Discovery and Explanation in Biology and Medicine, Chicago.

Schlimm, Dirk (2012): A new look at analogical reasoning, in: Metascience, 21 (1), pp. 197-201.

Schrader, Bernhard/Rademacher, Paul (2009): Kurzes Lehrbuch der Organischen Chemie, Berlin.

Shepherd, Gordon M. (2010): Creating modern neuroscience: The revolutionary 1950s, New York.

Silverstein, Arthur M. (1989): A History of Immunology, San Diego.

Silverstein, Arthur M. (2002): Paul Ehrlich's receptor immunology: The magnificent obsession, San Diego and London.

Strasser, Bruno (2006): A World in One Dimension: Linus Pauling, Francis Crick and the Central Dogma of Molecular Biology, in: History and Philosophy of the Life Sciences, 28, pp. 491-512.

Strasser, Bruno/Fantini, Bernadino (1998): Molecular Diseases and Diseased Molecules, in: History and Philosophy of the Life Sciences, 20 (2), pp. 189-214.

Travis, Anthony (1989): Science as a receptor of technology: Paul Ehrlich and the synthetic dye stuff industry, in: Science in context, 3 (2), pp. 383-408.

Travis, Anthony (1991): Paul Ehrlich: 100 years of chemotherapy, 1891-1991, in: The Biochemist, 13.

Travis, Anthony/Reinhardt, Carsten (2000): Heinrich Caro and the creation of modern chemical industry, Dordrecht.

Travis, Anthony (2008): Models for biological research: The theory and practice of Paul Ehrlich, in: History and Philosophy of the life sciences, 30, 79-98.

Weindling, Paul (1992): Scientific elites and laboratory organization in Fin-de-Siècle Paris and Berlin: The Pasteur Institute and Robert Koch Institute for Infectious Diseases compared, in: Cunningham/Williams (eds.): The laboratory revolution in medicine, Cambridge and New York, pp. 170-188.

Wimsatt (2007): Re-Engineering philosophy for limited beings, Cambridge (Ma).

Witkop, Bernhard (1999): Paul Ehrlich and his magic bullets –revisited, in: Proceedings of the American Philosophical Society, 143 (4), pp. 540-557.

Wolf, Gerhard (1970): Die BASF. Vom Werden eines Weltunternehmens, München.